Advances in Coastal Ecology - Processes, Patterns, and Services

Edited by Thomas Trott

Published in London, United Kingdom

Advances in Coastal Ecology - Processes, Patterns, and Services
http://dx.doi.org/10.5772/intechopen.1002160
Edited by Thomas Trott

Contributors
Abdulrahman Alsharhan, Bertha Jean Que, Christopher Kendall, Gerardo Rodríguez-Quiroz, Jerome Constantine Lekatompessy, Kannan Sivakumar, Kingsley John de los Santos, Mark S. Calabon, Parningotan Yosi Silalahi, Pedro Joaquín Gutiérrez-Yurrita, Perumal Murugesan, Rajagopal Gobalakrishnan, Resurreccion B. Sadaba, Sasmita Swain, Walter Hubbard-Zamudio

First published in London, United Kingdom, 2025 by IntechOpen
IntechOpen is the global imprint of INTECHOPEN LIMITED, registered in England and Wales, registration number: 11086078, 167-169 Great Portland Street, London, W1W 5PF, United Kingdom

For EU product safety concerns: IN TECH d.o.o., Prolaz Marije Krucifikse Kozulić 3, 51000 Rijeka, Croatia, info@intechopen.com or visit our website at intechopen.com.

British Library Cataloguing-in-Publication Data
A catalogue record for this book is available from the British Library

Advances in Coastal Ecology - Processes, Patterns, and Services
Edited by Thomas Trott
p. cm.

This title is part of the Environmental Sciences Book Series, Volume 25
Topic: Water Science
Series Editor: J. Kevin Summers
Topic Editor: Yizi Shang

Print ISBN 978-0-85014-841-1
Online ISBN 978-0-85014-840-4
eBook (PDF) ISBN 978-0-85014-842-8
ISSN 2754-6713

IntechOpen Book Series
Environmental Sciences
Volume 25

Aims and Scope of the Series

Scientists have long researched to understand the environment and man's place in it. The search for this knowledge grows in importance as rapid increases in population and economic development intensify humans' stresses on ecosystems. Fortunately, rapid increases in multiple scientific areas are advancing our understanding of environmental sciences. Breakthroughs in computing, molecular biology, ecology, and sustainability science are enhancing our ability to utilize environmental sciences to address real-world problems.

The four topics of this book series - Pollution; Environmental Resilience and Management; Ecosystems and Biodiversity; and Water Science - will address important areas of advancement in the environmental sciences. They will represent an excellent initial grouping of published works on these critical topics.

Meet the Series Editor

J. Kevin Summers is a Senior Research Ecologist at the Environmental Protection Agency's (EPA) Gulf Ecosystem Measurement and Modeling Division. He is currently working with colleagues in the Sustainable and Healthy Communities Program to develop an index of community resilience to natural hazards, an index of human well-being that can be linked to changes in the ecosystem, social and economic services, and a community sustainability tool for communities with populations under 40,000. He leads research efforts for indicator and indices development. Dr. Summers is a systems ecologist and began his career at the EPA in 1989 and has worked in various programs and capacities. This includes leading the National Coastal Assessment in collaboration with the Office of Water which culminated in the award-winning National Coastal Condition Report series (four volumes between 2001 and 2012), and which integrates water quality, sediment quality, habitat, and biological data to assess the ecosystem condition of the United States estuaries. He was acting National Program Director for Ecology for the EPA between 2004 and 2006. He has authored approximately 150 peer-reviewed journal articles, book chapters, and reports and has received many awards for technical accomplishments from the EPA and from outside of the agency. Dr. Summers holds a BA in Zoology and Psychology, an MA in Ecology, and Ph.D. in Systems Ecology/Biology.

Meet the Volume Editor

Thomas Trott was raised in a commercial fishing family a stone's throw away from the New England shore. He received degrees in Biology from the University of Massachusetts, Dartmouth (B.S.), Wake Forest University, North Carolina (M.A), and the Boston University Marine Program, located at the Marine Biological Laboratory, Woods Hole, Massachusetts (Ph.D.). His career highlights include being an advisor for citizen science programs at the New England Aquarium, Boston and the US EPA Superfund Program. Dr. Trott participated in the Census of Marine Life (2000-2010) as a PI in three of its programs: History of Marine Animal Populations, Natural History of Shore Areas, and Gulf of Maine Area Program. He developed and managed a benthic ecology program for the State of Maine Coastal Mapping Initiative. He is currently an Emeritus Professor of Biology at Suffolk University in Boston, where he taught ecology, biodiversity, and animal behavior for 19 years. An accomplished author of peer-reviewed publications, he serves as Editor, Guest Editor, and Associate Editor for scientific journals concerning marine biology, biodiversity, and natural history. Dr. Trott is particularly interested in assessing, documenting, and unravelling the influences of changes in fauna and flora on coastal processes and how these affect the quality of life in coastal communities.

Contents

Preface

Coastlines are subjected to a suite of complex interactions of natural processes originating from the clash of forces from the land and sea. They create a highly dynamic environment that shapes diverse habitats distributed along the world's shores. The dynamic interactions of water, sediments, and biota are intricate and can sometimes be seen on sandy shores and in exposed ancient seabeds. Coastal ecosystems provide multiple ecosystem services that benefit humankind in ways related to, for example, food production and shoreline protection [1]. However, the damaging consequences of human habitation, such as pollution, habitat destruction, and overfishing, threaten these benefits. These impacts often result in coastlines becoming more vulnerable to coastal storms and rising sea levels [2].

In summary, the benefits of coastal ecosystem services are balanced by the detriments of human coastal habitation. The destructive effects are unevenly distributed around the globe and often are greatest on highly populated coasts in nations with limited resources to reduce their impacts. Unfortunately, such regions are also areas where the biological communities inhabiting the shore are poorly known.

Plastic pollution is one of the greatest threats to the integrity of coastal ecosystems. From large debris to micro-sized particles, plastics have detrimental effects on sea life, ranging from entanglement of marine mammals of all kinds, turtles, and sea birds to ingestion of microplastics by suspension-feeding invertebrates like zooplankton, bivalves, and barnacles [3]. It accumulates in highly productive and sensitive habitats, damaging seagrass beds and altering the growth of mangroves, and in areas of high cover, causes death [4]. The history of plastic pollution, its composition, and its impacts are reviewed in Chapter 1, *Plastics in Coastal Ecosystems: Evaluation of Anthropogenic Marine Debris Pollution*. Public health implications are discussed, and management options are proposed. One effective way to manage plastic pollution is through community-led initiatives that remove debris while fostering environmental stewardship. A remarkable example that focuses on debris removal and long-term ecosystem rehabilitation in Indonesia is presented in Chapter 2, *Restoring Maluku's Coast: Action-Driven Beach Cleanups, Mangrove Planting, and Coral Renewal in Ambon by the Faculty of Medicine, University of Pattimura*. This curriculum-based program has successfully managed plastic waste that accumulates on the shores. Moreover, participants are restoring mangrove forests, which are vital to shoreline protection, and corals lost to sedimentation from land development and solid waste. The victories of this program motivate the creation of similar programs elsewhere around the globe.

Coastal environments are shaped by complex land and sea interactions and by the organisms that inhabit shorelines. Many species are ecosystem engineers and modify their surroundings by creating reefs, modifying sediments positively through feeding and bioturbation, and influencing nutrient cycling through feeding activities [5].

Polychaetes, known commonly as bristle worms or colloquially as sea worms, are important ecosystem engineers and modify sediments in the ways just mentioned, enhancing coastal ecosystem services by doing so. Chapter 3, *Ecological Interactions and Adaptations of Polychaetes in Coastal Habitats of East Coast of India*, reviews the diversity and ecology of polychaetes inhabiting the eastern coast of India. Their general biology, symbiotic relationships, and importance for bioremediation through bioturbation are described. Important aspects of polychaetes as biological indicators of pollution are highlighted, as are the human impacts on the diversity and abundance of bristle worms by coastal development, harvesting, and fish farming.

The significance of aquaculture for food production is undisputable, but the side effects of mariculture have been scrutinized with respect to their damaging effects on coastal ecosystems [6]. Impacts often cited are decreased water quality, degraded seabeds and biological communities beneath cages, loss of coastal habitats, transfer of disease, and reduced genetic fitness of wild conspecifics by escaped stocks. Chapter 4, *Perspective Chapter: Mixed Shrimp Farming and Mangrove Landscapes – between Myth and Reality*, presents a strategy for conserving mangrove forests as a highly productive shrimp mariculture industry in the Mexican State of Sinaloa grows. The proposed management plan relies on current landscape management concepts to develop bio-landscapes where shrimp farming and mangroves coexist.

The coastal microbiome of marine sediments strongly influences the productivity and services of coastal ecosystems. Marine microbes play a vital role in the biogeochemical cycles of coastal sediments through their conversion and mineralization of organic materials in sediments [7]. The marine actinobacteria are important constituents of the coastal microbiome. Chapter 5, *Ecological Factors Shaping Marine Actinobacterial Niches in Diverse Coastal Habitats of the Andamans Sea, India*, examines the physical and chemical factors that best define the realized niches of actinobacteria sampled from the field and cultured in the laboratory. A major outcome of this work is a framework for identifying the optimal habitats where the greatest abundance of actinobacteria can be found among different coastal habitats and where their contributions to ecosystem services could be the largest. Sometimes, the geochemical activity of the coastal microbiome leaves traces in ancient sediments that can inform the geological past of a coast. Chapter 6, *Cyanobacterial Mats in Stradline Holocene Carbonate: Evaporite Deposits along the Coast of Southern Arabian Gulf - Potential Sources of Petroleum in Ancient Carbonate Reservoirs*, examines the unusual coastal blue-green algal flats of the United Arab Emirates barrier islands. The processes uncovered by analysis of sediments and cyanobacteria suggest that oil fields in the Middle East, United States, and Canada may have originated from similar sediment-cyanobacteria associations. This work underscores the importance of coastal microbiomes and their significance beyond present-day coastal ecosystems.

This book benefited from the valuable contributions and timely correspondence of the authors, who are gratefully acknowledged. The excellent guidance of Nina Miocevic, Publishing Process Manager, IntechOpen, made the production process seamless. Finally, this book would not have been possible without the helpful assistance of Filip Lovricevic, Commissioning Editor at IntechOpen, in the early stages of its development.

Thomas J. Trott
Department of Biology,
Suffolk University,
Boston, United States of America

References

[1] Barbier EB, Hacker SD, Kennedy C, Koch EW, Stier AC, Silliman BR. The value of estuarine and coastal ecosystem services. Ecological monographs. May 2011;**81**(2):169-193

[2] d'Anna M, Idier D, Castelle B, Rohmer J, Cagigal L, Mendez FJ. Effects of stochastic wave forcing on probabilistic equilibrium shoreline response across the 21st century including sea-level rise. Coastal Engineering. 1 Aug 2022;**175**:104149

[3] Thushari GG, Senevirathna JD. Plastic pollution in the marine environment. Heliyon. 1 Aug 2020;**6**(8)

[4] van Bijsterveldt CE, van Wesenbeeck BK, Ramadhani S, Raven OV, van Gool FE, Pribadi R, et al. Does plastic waste kill mangroves? A field experiment to assess the impact of macro plastics on mangrove growth, stress response and survival. Science of the Total Environment. 20 Feb 2021;**756**:143826

[5] Passarelli C, Hubas C, Paterson DM. Mudflat ecosystem engineers and services. Mudflat Ecology. 2018:243-269

[6] Primavera JH. Overcoming the impacts of aquaculture on the coastal zone. Ocean & Coastal Management. 1 Jan 2006;**49**(9-10):531-545

[7] Underwood GJ, Dumbrell AJ, McGenity TJ, McKew BA, Whitby C. The microbiome of coastal sediments. The Marine Microbiome. 28 Feb 2022:479-534

Chapter 1

Plastics in Coastal Ecosystems: Evaluation of Anthropogenic Marine Debris Pollution

Kingsley John de los Santos, Mark S. Calabon and Resurreccion B. Sadaba

Abstract

Coastal ecosystems are often overlooked sinks of anthropogenic marine debris (AMD), persistent human-made materials with sizes ranging from 1 mm to 1 m which are mostly plastics. Although serving as transition zones between terrestrial and marine environments, the degree and extent of plastic pollution in coastal environments remains understudied. The chapter will discuss the background and history, composition, sources, and endpoints of AMD. Its spatiotemporal variabilities will be presented with case studies together with other related discoveries in coastal plastic pollution. The impacts of AMD pollution, specifically in mangroves and seagrass ecosystems, will be explored in depth. Discussions on public health implications, waste management initiatives, and future directions will conclude the chapter.

Keywords: marine debris, microplastics, coastal ecosystems, marine pollution, mangroves, seagrasses

1. Introduction

1.1 Overview of anthropogenic marine debris (AMD) and coastal ecosystems

In the past century, global plastic production has increased at an unprecedented rate, which has caused unforeseen impacts on the environment [1, 2]. With an estimated 40% of the global human population living in coastal areas, adjacent ecosystems are experiencing pressures that are only now beginning to be understood, particularly in relation to plastic pollution [3, 4]. Anthropogenic marine debris (AMD) is an example of human-produced macro debris washed off to marine and coastal habitats [5]. These debris can be ingested and/or entangled with various organisms and may have lethal or sub-lethal effects [6]. Only in the past decade have studies on AMD and its effects on marine and coastal ecosystems been recognized, as elevated densities of debris have been observed on mangroves and other coastal ecosystems [7]. The scope and severity of the impact of AMD on coastal ecosystems are just beginning to be understood. With more people living in the coastal areas of the globe, the world's coasts have never been more vulnerable.

1.2 Scope and objectives of the chapter

This chapter aims to provide a comprehensive overview of anthropogenic marine debris (AMD) in coastal ecosystems. The primary focus will be on the background and history, types, and endpoints of AMD, as well as its multifaceted impacts on coastal environments. Through detailed case studies, the chapter will examine the factors influencing spatiotemporal variations in AMD distribution and effects.

2. Background and history of AMD

AMD is any persistent human-generated macro material waste that ends up in marine and coastal environments [5]. The debris size threshold may vary, but the generally considered width size is within the range of 1–50 mm [8]. As an example of macro debris, AMD can fragment due to chemical and/or physical processes and may result in micro debris formation [9]. Historically, the majority of AMD is made of plastics that were manufactured commercially.

Plastic production began in 1907, but it was not until large-scale commercialization that production ramped up to around 20 million tonnes annually in the 1950s and 1960s. By the 1980s, global plastic production had surpassed 100 million tonnes and exponentially increased to nearly 500 million tonnes by 2019 (**Figure 1**) [2]. It is estimated that 79% of all the plastics ever produced and almost all of the plastic debris that has entered the ocean are still present [2, 10].

Debris in coastal ecosystems may vary in size from large, visible debris (macroplastics) to tiny fragments (microplastics). Macroplastics, such as discarded fishing gear, plastic bags, and bottles, can cause immediate physical damage to seagrass structures [6]. Microplastics, typically defined as particles smaller than 5 millimeters, originate from the breakdown of larger items and from direct sources like microbeads in personal care products [11]. These small particles are particularly concerning due

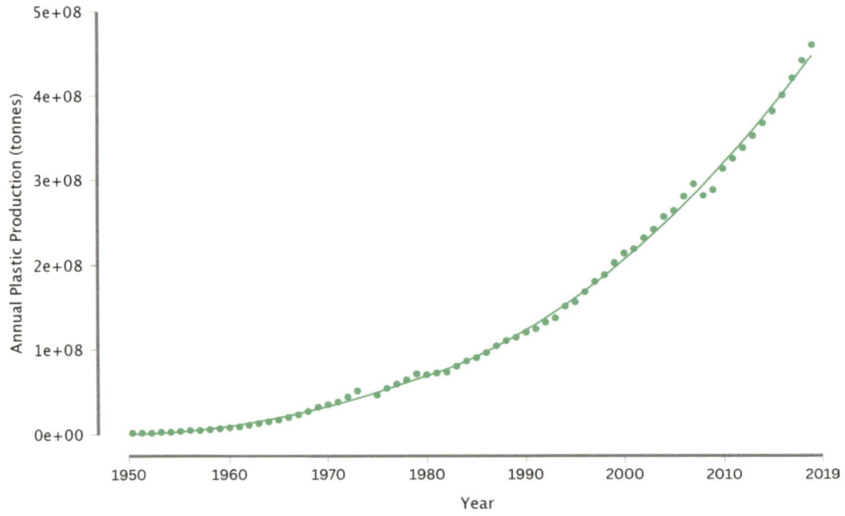

Figure 1.
Global annual plastic production from 1950 to 2019 [1, 2].

to their high abundance and potential for ingestion by marine organisms [12]. The sizes of plastics can influence their impact and distribution in the environment:

- Macroplastics (>25 mm): Macroplastics, such as single-use plastics like bottles and packaging, are highly visible and have garnered significant attention. Entanglement with seagrass leaves and rhizomes can cause damage, potentially leading to uprooting and smothering of the plants [6].

- Mesoplastics (25–5 mm): Mesoplastics fall between microplastics and macroplastics in size. Mesoplastics can have similar impacts to microplastics due to their high surface areas, especially absorption of leached compounds, ingestion, and bioaccumulation through food chain transversion [13].

- Microplastics (5 mm–0.1 μm): Microplastics are smaller plastic particles that can be ingested by a wide range of marine species, which have already been observed to enter the marine food webs throughout trophic levels. The ingestion of microplastics can impact the growth, survival, and reproduction of marine organisms, including those living in seagrass habitats [11].

- Nanoplastics (<0.1 μm): Nanoplastics are even smaller particles that can have diverse impacts on marine and coastal organisms. The main concern surrounding nanoplastics is their possible impact on human health as they accumulate on economically important fisheries products [14]. The pervasiveness of nano-sized plastics was further confirmed when their presence was recorded in human blood, raising public health concerns [15].

3. Composition and sources of AMD

3.1 Types of marine debris (plastic, metal, glass)

AMD can be of marine, coastal, or inland sources. The AMDs can be of various materials but are generally composed of plastic, polystyrene foam, glass, and metals [16, 17]. Fishing gear is also considered to be a major source of marine litter [18]. Sources of AMD can also vary and may come from inland as urban runoff and/or offshore origins (fisheries and aquaculture operations) [19, 20]. Coastal environments, especially mangrove areas, are sediment and material traps and may serve as long-term AMD stock and may experience reintroduction of AMD from the sea (**Figure 2**) [8, 21]. Large river deltas and tropical cyclones are also found on multiple occasions to flush significant amounts of inland-sourced AMD into the sea [22–27].

As much as 80% of macro-marine debris is plastic [16]. Plastics can be generally classified into sizes and chemical compositions. Size classes in width include mega plastic (>1 m), macro plastic (10 mm–1 m), meso-plastic (10 mm–1 mm), and microplastic (<1 mm–1 μm) [28, 29]. The anthropogenic marine debris (1–50 mm) includes all meso- and macro plastics, and a portion of mega plastic size class. In addition, the chemical composition of plastics influences their respective densities. Polypropylene (PP) and polyethylene (PE) float and are easily dispersed, while polystyrene (PS), polyvinyl chloride (PVC), polyamide (PA), and polyethylene terephthalate (PET) are more likely to be suspended in the water column and get deposited on the seafloor [30].

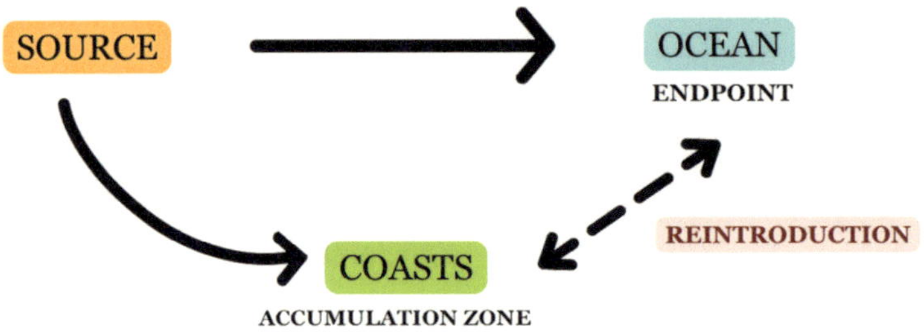

Figure 2.
Schematic diagram of reintroduction of AMD between plastic coastal and oceanic zones.

Among the plastics found in seagrass meadows, certain polymers are more prevalent. Polyethylene (PE) and polypropylene (PP), which are used in a wide range of consumer products, are commonly identified. Polystyrene (PS) and polyvinyl chloride (PVC) are also frequently found, along with less common polymers like nylon and polyester. The chemical properties of these polymers influence their degradation rates and interaction with the marine environment [2].

3.2 Types of plastics according to chemical structure

The types of plastics in terms of chemical composition are coded with numbers 1–7 [31].

3.2.1 Polyethylene terephthalate (PET) (type 1)

PET is commonly used for items like soda bottles and food jars and containers.

3.2.2 High-density polyethylene (HDPE) (type 2)

HDPE is found in products like milk jugs and detergent bottles. HDPE is often used in for food packaging due its heat tolerance.

3.2.3 Polyvinyl chloride (PVC) (type 3)

PVC is often used in products like pipes, clear food packaging, and shrink wrap due to its durability.

3.2.4 Low-density polyethylene (LDPE) (type 4)

LDPE is commonly found in products like plastic bags, six-pack rings, and various containers. LDPE is relatively safe but can degrade under prolonged exposure to sunlight and heat.

3.2.5 Polypropylene (PP) (type 5)

PP is usually used in products like straws, bottle caps, and food containers with minimal risk of leaching.

3.2.6 Polystyrene (PS) (type 6)

PS is found in packing peanuts, *Styrofoam* cups, and food containers. Due to its light weight and insulation properties, it is often used in food packaging and insulation materials.

3.2.7 Other plastics (type 7)

Plastic marked with a 7 includes materials other than those listed in numbers 1–6.

These different types of plastics can enter seagrass beds, posing a threat to these vital marine ecosystems. Proper waste management and responsible plastic use are essential to prevent plastic pollution from harming seagrass meadows.

3.3 Sources and endpoints of AMD

It is estimated that only 20 rivers are contributing to two-thirds of total plastic entering the oceans globally [32]. Fishing, aquaculture, shipping, and sea-based mining are also major contributors to marine and coastal macro debris pollution [33–36]. Other land-based mechanisms of AMD introduction in the coastal and marine environment may be through (a) urban runoff and stormwater systems, (b) rivers and streams, (c) residential and domestic activities, (d) industrial and commercial sources, (e) agricultural activities, (f) tourism and recreational activities, and (g) mismanaged waste in urban centers [37].

While plastic pollution originating from land-based activities holds a prominent position in marine environments, it's essential to recognize the significant contribution of sea-based sources. These sources exacerbate the pervasive issue of marine plastic pollution and encompass:

3.3.1 Fishing gear and equipment

The issue of lost or abandoned fishing gear, commonly referred to as "ghost gears," represents a significant menace to marine ecosystems. These remnants of fishing activities, including nets, lines, traps, and various equipment, endure in the oceanic environment long after their initial deployment. They act as silent predators, ensnaring marine life indiscriminately, from small fish to large mammals like whales [18].

3.3.2 Aquaculture operations

The aquaculture industry is essential to satisfying the world's increasing demand for seafood. Although aquaculture plays a crucial role in food production, it also contributes significantly—though less visibly—to marine plastic waste. Plastic-based equipment, like nets, ropes, buoys, and containment systems, is widely used in aquaculture operations. While these materials are necessary for fish farming, their synthetic nature makes them brittle and easily detached, especially in unfavorable weather conditions. Plastic waste from aquaculture activities enters marine environments during storm events or as a result of insufficient containment mechanisms. These pieces, which may be anything from tiny particles to large equipment components, get into coastal seas and other areas, making the already widespread problem of marine plastic pollution worse.

3.3.3 Shipping and boating activities

The domains of recreational boating and commercial shipping, which are essential to both international trade and leisure pursuits, contain a substantial but sometimes disregarded source of marine plastic pollution. Commercial shipping boats play a crucial role in the world's maritime commerce networks, facilitating the movement of products over large regions of the ocean. Additionally, operational practices, including routine maintenance and cleaning procedures, may result in the discharge of plastic debris into marine waters, particularly in regions lacking stringent regulatory oversight. The accumulation of plastic debris from shipping and boating activities poses multifaceted threats to marine life, habitats, and human well-being. Marine organisms, ranging from microscopic plankton to charismatic megafauna, face the risk of ingestion or entanglement in plastic debris, leading to adverse health effects and mortality. Moreover, the persistence of plastic pollutants in marine environments undermines ecosystem resilience, disrupts natural processes, and jeopardizes the provision of essential ecosystem services [38].

3.4 Accumulation zones, coastal and oceanic sinks

The origin of the AMD may come from land-based or sea-based sources, but the route ultimately ends up in the oceans. Once released into the water column, high-density AMD will often sink and accumulate in areas where it was released [22, 39, 40] while the majority of AMD that is made with low-density materials such as polyethylene and polystyrene foam floats and follows sea surface currents [41, 42]. Floating marine debris may break into smaller pieces (e.g., microplastics) due to chemical and physical stresses and can affect a wide array of marine organisms [43]. Low-density AMD will eventually sink or be deposited on the shore, while some may remain afloat in one of the oceanic gyres and be a persistent part of the water column [38, 40]. Deposited AMD on the shore may be reintroduced to the ocean or become part of the "missing plastics" or the unaccounted in-ocean plastics from the ocean input estimates (**Figures 2–3**).

With coastal ecosystems being suspected traps of marine plastics, mangroves are proposed to be one of the primary traps [8]. Mangroves being a transition boundary between inland and oceanic sources can create areas of high AMD density. The global average of the number of AMD is 573 n km^{-2} but is dwarfed by AMD estimates in mangrove areas, which range from 54 n km^{-2} [47] to 21,230 n km^{-2} [7]. Although the global average for the number of AMD in mangrove areas is yet to be determined, the values provide insight into the degree and extent of AMD in these overlooked ecosystems. Estimates of plastics found in the open ocean are orders of magnitude lower than the estimates introduced from inland sources [45]. This suggests that there can be understudied areas in which plastics can accumulate—mangroves are one of the possible sinks of this debris. The mechanical complexity of the root systems provides a perfect biological filter for debris such as sediments and plastic debris [47, 48].

3.5 Long-term persistence and degradation

Plastics accumulate in the environment primarily due to their durability and resistance to natural degradation processes. When plastic waste enters natural environments, it begins to break down through physical, chemical, and biological processes:

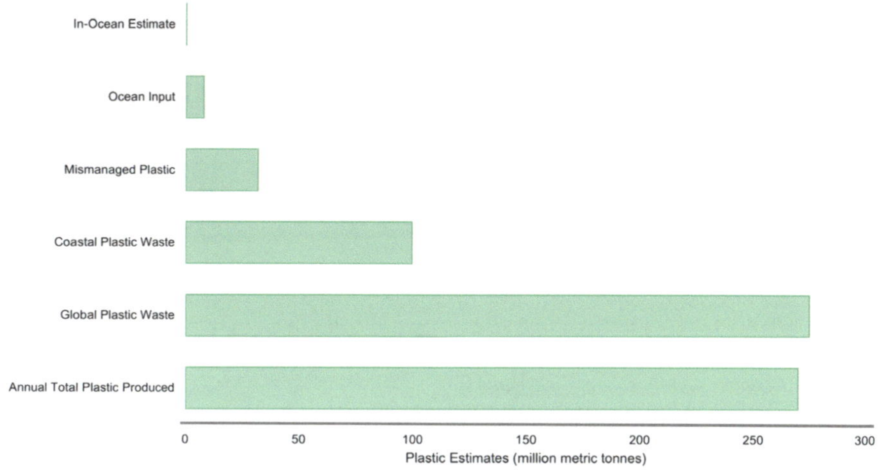

Figure 3.
The Missing Plastic Dilemma *shows the underestimation of plastics in the world's oceans (million metric tonnes) adapted from [44] based on [45, 46].*

3.5.1 Physical fragmentation

Larger plastic items (macroplastics and mesoplastics) exposed to sunlight (UV radiation), wave action, and mechanical forces fragment into smaller pieces, eventually forming microplastics (less than 5 mm in size).

3.5.2 Chemical degradation

Plastics can undergo chemical reactions that break down polymer chains, although this process is slow and often incomplete in natural environments.

3.5.3 Biological degradation

Certain microorganisms can degrade some types of plastics, but this process is generally very slow and limited in scope.

The persistence of plastics in the environment has led to their accumulation in a variety of ecosystems, including oceans, rivers, soils, and even the atmosphere. The persistence and pervasiveness of plastics are exceptionally demonstrated by its presence in Arctic ice, remote islands, and deep-sea sediments [40, 49–52].

4. Spatiotemporal variabilities in debris densities

4.1 Role of physical and chemical properties of plastics in marine environment

Density is the primary driver of distribution patterns of debris in aquatic ecosystems. With 80% of debris composed of plastics, the type of plastic is an important factor in where plastics end up [16].

Plastics in marine and coastal environments can undergo physical weathering and chemical degradation through sun exposure, biofilm growth, and oxidation [38]. The density of the plastics relative to water (1.00 g cm^{-3}) also determines whether

Type	Plastic type	Chemical formula	Density (g cm^{-3})	Source
1	PET	$(C_{10}H_8O_4)_n$	1.30–1.50	[53]
2	HDPE	$(C_2H_4)_n$	1.00	[54]
3	PVC	C_2H_3Cl	1.44	[54]
4	LDPE	$(C_2H_4)_n$	0.97	[54]
5	PP	$(C_3H_6)_n$	0.88–1.23	[53]
6	PS	$(C_8H_8)_n$	1.04–1.50	[53]
7	Other Plastics: PE	$(C_2H_4)_n$	0.92–0.97	[53]

Table 1.
Different kinds of plastics and their respective densities.

the plastics will remain at the surface or sink (**Table 1**). Properties of plastics that may undergo change due to extended exposure to the environment are color, surface characteristics, crystallinity, size, and density [55, 56].

4.2 Case studies highlighting spatiotemporal variability

4.2.1 Case study 1: Microplastics in equatorial coasts: Pollution hotspots and spatiotemporal variations associated with tropical monsoons

This study provides the first set of seasonal data on coastal marine plastic pollution in the equatorial region, highlighting the significant influence of monsoons. The spatiotemporal variations in marine plastic concentrations are strongly correlated with monsoonal patterns, particularly the high wind speeds and rainfall typical during these seasons. High wind speeds reintroduce and re-entrain plastic debris into the marine environment, while heavy rainfall washes terrestrial plastics into coastal waters. Overall, monsoon-driven wind and rainfall affect the dynamics of marine plastic pollution. Additionally, elevated concentrations of marine plastics were also observed in mangrove ecosystems, aligning with findings from other studies [48, 52, 57]. Mangroves, with their complex root systems and sheltered nature, act as significant traps for marine debris. This monsoon interaction with the spatiotemporal variabilities creates more opportunities in research to understand the direct interplay among plastic pollution, hydrogeography, weather, and indirectly through public health and management [58].

4.2.2 Case study 2: Accumulation and exposure classifications of plastics in the different coastal habitats in the western Philippine archipelago

Although the Philippines has consistently ranked as one of the top contributors to marine plastic pollution, spatiotemporal factors to these variabilities of plastics in the country remain unclear. This study has identified three potential key factors in plastic accumulation, namely [59]:

4.2.2.1 Population density

The study suggests that highly populated and urbanized areas are related to single-use plastic, which in turn results in higher plastic waste in adjacent coastal ecosystems. These heightened levels of plastic also create a cascade of issues, such as breeding grounds for vectors of diseases and flooding [60].

4.2.2.2 Wind and oceanic transport

Interestingly, the presence of comparable debris density in remote and rural coastal areas with urban areas has suggested that hydrography—sea surface currents—is also a key influencer in the pollution level of the areas.

4.2.2.3 Habitat type

The study assessed the average total plastic pieces per coastal ecosystem: seagrasses (8 sites; 12.58 ± 18.55 n), mangroves (8 sites; 530.13 ± 955.94 n), and beaches (13 sites; 288.41 ± 380.99 n). Although no significant differences were observed between mangroves and beaches, seagrasses were found to have significantly lower plastic density. This suggests that mangroves' root system complexity may have played a role in the retention of debris, a finding that is consistent globally.

5. Impact of AMD on coastal environments

5.1 General environmental consequences

5.1.1 Impact on biodiversity and introduction of nonnative species

The spread of plastic pollution poses a grave danger to marine biodiversity, with an alarming 2150 marine species encountering plastic debris in their habitats [61]. Tragically, these encounters often lead to entanglement, ingestion, and habitat alteration, imperiling the very fabric of marine life. The presence of plastics reverberates through ecosystems, disrupting vital species interactions, altering population dynamics, and reshaping community structures. The net result is a cascade of biodiversity loss and ecosystem simplification, threatening the resilience and stability of coastal environments [52]. One intriguing impact of AMD on biodiversity is their role as vectors of invasive species. As these debris get caught with the currents, arthropods may use these as rafts and be a medium for the introduction of nonnative and even invasive species [62]. This risk is further exacerbated in insular ecosystems in which island-dwelling species are especially vulnerable to the encroachment of alien species [63].

5.1.2 Impact on trophic levels

Plastic pollution's ecological ramifications extend beyond mere biodiversity loss, infiltrating trophic levels, and causing further complications through biomagnification and/or bioaccumulation [64, 65]. Emerging studies reveal a disturbing trend: species loss disproportionately affects large consumers, triggering profound shifts in trophic structure and ecosystem functioning [66]. Effects of trophic disruption can be overserved from primary producers such as phytoplankton to organisms in high trophic levels—marine mammals and fish [12, 67]. As debris moves through the trophic levels, potential implications for human health are considered [64, 67].

5.2 Ecosystem-specific impacts

5.2.1 Mangroves

Mangroves are a group of halophytic intertidal plants distributed along tropical and subtropical shorelines of the world [68]. Globally, these forests cover an estimated 137,760 km^2 or 0.01% of the total land area of the planet [69]. Even with this minimal cover, mangroves provide an invaluable ecosystem. They serve as carbon sinks as much as five-fold relative to terrestrial forests and breeding grounds and nurseries for diverse sets of organisms [70, 71].

Industrialization and fisheries (mainly aquaculture), the two most cited causes of mangrove deforestation, have resulted in increasing waste released into the bodies of water and created new and often overlooked problems: AMD and microplastics [72, 73]. Industrialization, especially in nearshore areas, often results in bottles, bags, resin, and packaging being released into the ocean [72]. Lost and abandoned fishing gear, mostly made of plastic materials such as nylon and synthetic polymers, are also contributors to anthropogenic marine debris [74].

The amount of time it takes to biodegrade plastics and their buoyancy—especially polyethylene and propylene used in packaging—contribute to their tremendous dispersal potential [75]. The dispersion is, in turn, influenced by wind and oceanic currents [76]. The released plastic can ultimately end up in oceanic gyres, deposited on the ocean floor, or trapped in mangroves. The last of which is still poorly understood despite the services provided by mangrove ecosystems.

A World Bank Group reports that litter concentrations in the Caribbean are generally higher in areas that are close to high populations coupled with poor solid waste management infrastructure [77]. This is consistent with findings across different regions—the Mediterranean and Southeast Asia [78–80].

It was only in 2010 that studies on mangrove AMD gained attention (**Table 2**). It can be observed that mangroves (21.23 n m^{-2}) have higher AMD densities as compared with the beach and open shore sites (5.59 n m^{-2}) [7]. Higher densities were reported in an Indonesian study with 533 n per m^{-2} (**Table 2**) [90].

5.2.2 Seagrass

Seagrass meadows are highly productive marine ecosystems that provide critical habitat and ecosystem services, but they are increasingly threatened by plastic pollution. Factors such as water currents, wave action, and density and structure of seagrass meadows make them susceptible to debris accumulation. The physical structure of seagrass blades can capture floating plastics and debris, preventing them from dispersing further into the ocean. This trapping mechanism leads to the accumulation of plastics in seagrass beds, where they can persist for extended periods. In the Mediterranean Sea, a study found that seagrasses can accumulate up to 650 macroplastic items per square meter, with the majority classified as plastic bags and food packaging [22]. Another study in Australia reported that seagrass beds can trap up to 80% of the plastic debris in coastal areas [91]. In terms of microplastics, these ecosystems can also harbor hundreds of pieces (884.6 n kg^{-1}) highlighting their role with mangroves as overlooked sinks of marine debris. This increased level of debris relative to other ecosystems may account for *The Missing Plastic Dilemma* as depicted in **Figure 3** [92].

Site	Method	Average items per m^2	Range of no. of items per m^2	Debris size threshold (width)	Year	Source
Kingston, Jamaica	$1m^2$ quadrats, transect line parallel to the shore	n/a	n/a	>2 mm	1996	[81]
Sao Paulo, Brazil	$10m^2$	1.33	n/a	n/a	2010	[82]
Goiana Estuary, Brazil	Cylindrical sediment coring (20 cm dia, 20 cm ht)	n/a	n/a	>1 mm	2011	[83]
Mutapore Island, Papua New Guinea	1 m x 2 m plots	21.23	1.2–78.3	>5 mm	2012	[7, 84]
Mumbai, India	n/a	n/a	n/a	n/a	n/a	[84]
Lac Bay, Bonaire Island, Caribbean	Belt-Transect, 5 m wide (perpendicular to the shore)	11.76	5.8–23.2	>50 mm	2013	[85]
Santa Catarina Island, Caribbean	50 m^2	n/a	n/a	n/a	2013	[86]
Pantai Indah Kapuk, Jakarta, Indonesia	Multiple 1 m^2 quadrats	n/a	20–533	>2 mm	2014	[87]
Tunda Island, Indonesia	100 m x 1 m transect	0.75	n/a	n/a	2018	[48]
Saudi Arabia	Belt-transect between 2 and 8 m wide and 4–60 m long	0.66	0.02–3.7	>25 mm	2019	[48]
Buenaventuar, Colombia	154 m^2	9.42	0.22–35.4	>5 mm	2019	[88]
Ciénaga Grande de Santa Marta, Colombia	Belt Transects 50 m x 2 m, perpendicular to the shore	0.054	0.015–0.073	>25 mm	2019	[47]
Penang Island, Malaysia	10 m x 10 m quadrats along a transect line parallel to the shore	n/a	2.15–73.12	>25 mm	2020	[89]

Table 2.
Summary of mangrove macro debris studies. n/a indicates that the data is not reported/sampled [8].

5.2.2.1 Physical impacts

5.2.2.1.1 Smothering and abrasion of seagrass blades

Plastics, especially larger debris such as bags and fishing nets, can smother seagrass blades, blocking essential gas exchange and nutrient absorption. The weight and movement of these plastics can cause abrasion and physical damage to the delicate structures of the seagrass, leading to tissue damage, reduced growth rates, and even mortality of the plants [93].

5.2.2.1.2 Changes in light penetration and sediment characteristics

On top of physical smothering, light penetration can also be affected by increased density of suspended debris, which can affect photosynthesis [91]. This shading inhibits the ability of seagrasses to capture sunlight and convert it into energy through photosynthesis, thereby limiting their growth and productivity. Additionally, plastics can change the physical characteristics of sediments by increasing sediment compaction and altering grain size distribution [94]. As plastics accumulate on the seafloor, they can compact the sediment layer, reducing pore space and limiting water and nutrient exchange. Alterations in grain size distribution can also affect sediment stability and permeability, further impacting the availability of oxygen and nutrients and the anchorage of the seagrass roots.

5.2.2.2 Biological impacts

Seagrass beds are beginning to be understood as sinks for marine plastics [52, 92]. Although marine plastics are ubiquitous in coastal ecosystems, especially seagrasses, only 15% of seagrass-associated species have been studied for marine plastics [95]. In a review last 2023, the average microplastics in seagrass meadows is 0.085 ± 0.031 n/cm^2 based on 37 studies [95]. It was noted that the presence of epiphytes and biofilms increased the chances of seagrass trapping microplastics, with the highest abundance found in *Thalassia hemprichii* [95]. In terms of vulnerability, the main threat to seagrass beds comes from the physical abrasion and reduction in light energy brought about by the large debris [96]. In terms of microplastics, it is often the seagrass-associated fauna that suffers the consequences [96]. The highest abundance of plastics was found in sea cucumbers with 52 ± 7.68 n/individual and commercial bivalves at 57.2 n/individual [97]. In terms of impact, leached plastics have been observed to affect the embryonic development of sea urchins (*Lytechinus variegatus*) [98]. Although still limited, the impact of plastics already points to the direction of movement of plastics within the trophic levels [66]. With its detection of commercial fishery products, the plastic pollution concerns transcend the seagrass beds and extend to public health.

6. Public health implications

Marine debris, particularly plastics, pose significant direct health risks to human populations in coastal areas. Studies in China and Southeast Asia have found high levels of microplastics in commercially important fish species, posing a significant risk for consumers [99, 100]. Communities residing in these regions often source their nutrition needs primarily through capture fisheries, aquaculture, and/or gleaning. The ingestion of microplastics and associated toxins by marine organisms can lead to bioaccumulation and biomagnification of toxins, such as heavy metals and persistent organic pollutants (POPs), in the human food chain [101]. Consuming contaminated seafood can result in adverse health effects, including gastrointestinal issues, endocrine disruption, and an increased risk of cancer [99, 102].

Plastic debris in estuarine environments can also affect the health of humans in indirect ways, which can lead to waterborne diseases. During floods or high tides, contaminated plastics can spread pathogens to residential areas, increasing the incidence of diseases such as cholera, typhoid, and hepatitis [6]. Additionally, debris in urban areas can trap water, which serves as breeding sites for disease-carrying vectors like mosquitoes, contributing to outbreaks of vector-borne diseases such as dengue fever and malaria [60]. High rainfall during monsoons was also found to positively correlate with plastic density, further increasing the risk brought by the urban plastics in urban tropical areas [58]. The realm of plastic pollution and infectious disease is complex and interdisciplinary and is just beginning to be understood, may it be in coastal or urban settings.

7. Future directions in AMD management

7.1 Advanced waste management

Advances in integrated waste management approaches combine various methods, including recycling, composting, and waste-to-energy processes, to minimize waste generation and improve disposal efficiency in recent years. Technologies such as automated sorting systems use optical sensors and artificial intelligence (AI) to sort waste from recycling facilities in countries like Japan and South Korea [103]. Advances in waste management include plasma gasification, which uses high-temperature plasma (> 6000°C). This converts the waste to synthesis gas (syngas), which generates energy. This technology is being piloted in several countries, such as Singapore and China, as part of their advances in waste management strategies [104].

7.2 Managing AMD through community-led initiatives

Community-led management of AMDs is increasingly popular in creating more sustainable solutions for mitigating human impact on coastal ecosystems. Programs like the *International Coastal Cleanup* have mobilized thousands of volunteers in countries such as the Philippines and Indonesia, effectively removing debris from shorelines and raising awareness about marine pollution [105]. These clean-ups address immediate litter issues and foster environmental stewardship among participants. Education and awareness campaigns, such as those conducted by *Clean Up the World*, play a crucial role in encouraging sustainable practices and reducing plastic use [106].

Local policy advocacy and innovative solutions tailored to specific contexts enhance AMD management further. In Indonesia, community-driven efforts have established localized recycling centers and eco-friendly alternatives to plastic products. The *Plastic Bank* initiative incentivizes waste collection by allowing community members to exchange plastic waste for goods or cash, providing economic benefits while addressing pollution [107]. Citizen science projects, like those facilitated by the *Marine Debris Tracker* app, enable residents to contribute valuable data on marine debris, supporting broader environmental monitoring and policy development efforts [108]. These grassroots initiatives leverage local knowledge and foster collaborations with governments and non-government organizations, playing a pivotal role in combating marine pollution sustainably.

7.3 Conclusion

The impact of AMD, particularly plastics, on coastal ecosystems is only beginning to be understood. With the increase in the body of knowledge comes more questions about the dynamics, distribution, and long-term effects of the ever-increasing plastic pollution in the world's coastal environments. One key challenge is the need for standardized methods to assess marine debris, enhancing data comparability for more effective monitoring. Innovations in waste management—such as leveraging citizen science, utilizing machine learning to model debris distribution, and exploring the socioeconomic drivers of pollution in coastal communities—could be crucial in addressing this growing issue. Furthermore, investigating how plastics interact with contaminants and assessing their broader impacts on ecosystem services, like carbon sequestration and coastal protection, could reveal possible links to climate change.

Plastics, with their ecological impacts, continue to shape the world in unprecedented ways. Threats caused by its pervasiveness in various ecosystems, while continuing to open opportunities for further research, also reveal the vulnerabilities of marine and coastal environments. Upon its invention, plastics were deemed a breakthrough for their lightweight, durable, and cost-effective properties, making them a revolutionary alternative to traditional materials like wood and metal. Ironically, these very characteristics contribute to today's environmental crises. What was seen as a solution to day-to-day problems decades ago became a major problem that continues to persist. The increasing understanding of plastic pollution underscores a shared global responsibility to advance not only scientific research but also policy and management strategies that promote a healthier and more sustainable future. Like other global challenges—such as ozone depletion and climate change—plastic pollution reveals the ecosystem complexities, often overlooked in profit-driven and reductionist paradigms. Beyond the scientific concepts presented lie the lives and societies deeply affected by this systemic issue. Acknowledging individual responsibility while highlighting the role government institutions and corporations can play in driving real change by establishing accessible, effective waste management systems and ultimately addressing the environmental and societal consequences of plastic pollution.

Acknowledgements

The authors would like to acknowledge the University of the Philippines Visayas for the funding of this publication.

Conflict of interest

The authors declare no conflict of interest.

Author details

Kingsley John de los Santos*, Mark S. Calabon and Resurreccion B. Sadaba
University of the Philippines Visayas, Iloilo, Philippines

*Address all correspondence to: kdelossantos@up.edu.ph

References

[1] Global Plastics Outlook. OECD; 2022 [cited 2024 Jul 12]. Available from: https://www.oecd-ilibrary.org/environment/global-plastics-outlook_de747aef-en

[2] Geyer R, Jambeck JR, Law KL. Production, use, and fate of all plastics ever made. Science Advances. 2017;**3**(7) Available from: https://plasticoceans.org/wp-content/uploads/2018/05/Production_use_and_fate_of_all_plastics_ever_made.pdf

[3] Reimann L, Vafeidis AT, Honsel LE. Population development as a driver of coastal risk: Current trends and future pathways. Cambridge Prisms: Coastal Futures. 2023;**1**:e14. Available from: https://www.cambridge.org/core/product/8261D3B34F6114EA0999FAA597D5F2E2

[4] Horton AA. Plastic pollution: When do we know enough? Journal of Hazardous Materials. 2022;**422**:126885. Available from: https://www.sciencedirect.com/science/article/pii/S0304389421018537

[5] Coe JM, Rogers DB. Marine Debris. In: Coe JM, Rogers DB, editors. Springer Series on Environmental Management. New York: Springer; 1997. pp. 161-170

[6] Gall SC, Thompson RC. The impact of debris on marine life. Marine Pollution Bulletin. 2015;**92**(1-2):170-179

[7] Smith SDA. Marine debris: A proximate threat to marine sustainability in Bootless Bay, Papua New Guinea. Marine Pollution Bulletin. 2012;**64**(9):1880-1883

[8] Luo YY, Not C, Cannicci S. Mangroves as unique but understudied traps for anthropogenic marine debris: A review of present information and the way forward. Environmental Pollution. 2021;**271**:116291

[9] Hurley RR, Nizzetto L. Fate and occurrence of micro(nano)plastics in soils: Knowledge gaps and possible risks. Current Opinion in Environmental Science & Health. 2018;**1**:6-11

[10] Thompson R. New directions in plastic debris. Science (1979). 2005;**310**(5751):1117b-1117b

[11] Cole M, Lindeque P, Halsband C, Galloway TS. Microplastics as contaminants in the marine environment: A review. Marine Pollution Bulletin. 2011;**62**(12):2588-2597

[12] Wright SL, Thompson RC, Galloway TS. The physical impacts of microplastics on marine organisms: A review. Environmental Pollution. 2013;**178**:483-492. Available from: https://www.sciencedirect.com/science/article/pii/S0269749113001140

[13] Devi SS, Kumar AB. Impacts of mesoplastics and microplastics on marine biodiversity. In: Sekar S, Senapathi V, Sabarathinam C, Viswanathan PM, editors. Meso and Microplastic Risk Assessment in Marine Environments. Elsevier; 2024. pp. 309-340. Available from: https://www.sciencedirect.com/science/article/pii/B9780323909808000145

[14] Zaki MRM, Aris AZ. An overview of the effects of nanoplastics on marine organisms. Science of the Total Environment. 2022;**831**:154757. Available from: https://www.sciencedirect.com/science/article/pii/S0048969722018502

[15] Leslie HA, van Velzen MJM, Brandsma SH, Vethaak AD, Garcia-Vallejo JJ, Lamoree MH. Discovery and quantification of plastic particle pollution in human blood. Environment International. 2022 May;**163**:107199

[16] Derraik JGB. The pollution of the marine environment by plastic debris: A review. Marine Pollution Bulletin. 2002;**44**(9):842-852

[17] Lee J, Hong S, Song YK, Hong SH, Jang YC, Jang M, et al. Relationships among the abundances of plastic debris in different size classes on beaches in South Korea. Marine Pollution Bulletin. 2013;**77**(1-2):349-354

[18] Macfadyen G, Huntington T, Cappell R, Nations FAAOOTU, Programme UNE. Abandoned, Lost or Otherwise Discarded Fishing Gear. Rome, Italy: United Nations Environment Programme; 2009

[19] Kiessling I. Finding Solutions: Derelict Fishing Gear and Other Marine Debris in Northern Australia. Hobart, TAS (Australia): National Oceans Office; 2003

[20] Otley H, Ingham R. Marine debris surveys at Volunteer Beach, Falkland Islands, during the summer of 2001/02. Marine Pollution Bulletin. 2003;**46**(12):1534-1539

[21] Nagelkerken I, Wiltjer GAMT, Debrot AO, Pors LPJJ. Baseline study of submerged marine debris at beaches in Curaçao, West Indies. Marine Pollution Bulletin. 2001;**42**(9):786-789

[22] Galgani F, Leaute JP, Moguedet P, Souplet A, Verin Y, Carpentier A, et al. Litter on the sea floor along European coasts. Marine Pollution Bulletin. 2000;**40**(6):516-527

[23] Cunningham DJ, Wilson SP. Marine debris on beaches of the greater Sydney region. Journal of Coastal Research. 2003;**19**(2):421-430. Available from: http://www.jstor.org/stable/4299182

[24] Abu-Hilal AH, Al-Najjar T. Litter pollution on the Jordanian shores of the Gulf of Aqaba (Red Sea). Marine Environmental Research. 2004;**58**(1):39-63

[25] Ribic CA, Sheavly SB, Klavitter J. Baseline for beached marine debris on Sand Island, midway atoll. Marine Pollution Bulletin. 2012;**64**(8):1726-1729

[26] Williams A, Simmons S. The degradation of plastic litter in rivers: Implications for beaches. Journal of Coastal Conservation. 1996;**2**:63-72

[27] Araújo MC, Costa M. An analysis of the riverine contribution to the solid wastes contamination of an isolated beach at the Brazilian northeast. Management of Environmental Quality: An International Journal. 2007;**18**(1):6-12

[28] Stubbins A, Law KL, Muñoz SE, Bianchi TS, Zhu L. Plastics in the earth system. Science (1979). 2021;**373**(6550):51-55

[29] Browne MA, Galloway TS, Thompson RC. Spatial patterns of plastic debris along estuarine shorelines. Environmental Science & Technology. 2010;**44**(9):3404-3409

[30] Guo X, Wang J. The chemical behaviors of microplastics in marine environment: A review. Marine Pollution Bulletin. 2019;**142**:1-14

[31] Thompson RC, Swan SH, Moore CJ, vom Saal FS. Our plastic age. Philosophical Transactions of the

Royal Society B: Biological Sciences. 2009;**364**(1526):1973-1976

[32] Lebreton LCM, van der Zwet J, Damsteeg JW, Slat B, Andrady A, Reisser J. River plastic emissions to the world's oceans. Nature Communications. 2017;**8**(15611):15611. Available from: https://www.nature.com/articles/ncomms15611

[33] Dixon TJ, Dixon TR. Marine litter distribution and composition in the North Sea. Marine Pollution Bulletin. 1983;**14**(4):145-148

[34] Schrey E, Vauk GJM. Records of entangled gannets (Sula bassana) at Helgoland, German bight. Marine Pollution Bulletin. 1987;**18**(6):350-352

[35] Dufault S, Whitehead H. Floating marine pollution in 'the gully' on the continental slope, Nova Scotia, Canada. Marine Pollution Bulletin. 1994;**28**(8):489-493

[36] Hinojosa IA, Thiel M. Floating marine debris in fjords, gulfs and channels of southern Chile. Marine Pollution Bulletin. 2009;**58**(3):341-350. Available from: https://www.sciencedirect.com/science/article/pii/S0025326X08005055?casa_token=hKmpYgbCy_YAAAAA:4fmBeMGHj_RmKU1ZKH-tpHatwTjtFKGHdooBNLt7uBzFul0b6GHzgiKq2WVNiCbtbz6oS0Y

[37] Thushari GGN, Senevirathna JDM. Plastic pollution in the marine environment. Heliyon. 2020;**6**(8):e04709. Available from: https://www.sciencedirect.com/science/article/pii/S2405844020315528

[38] Andrady AL. Microplastics in the marine environment. Marine Pollution Bulletin. 2011;**62**(8):1596-1605. Available from: https://www.sciencedirect.com/science/article/pii/S0025326X11003055

[39] Barnes DKA, Galgani F, Thompson RC, Barlaz M. Accumulation and fragmentation of plastic debris in global environments. Philosophical Transactions of the Royal Society B: Biological Sciences. 2009;**364**(1526):1985-1998

[40] Mordecai G, Tyler PA, Masson DG, Huvenne VAI. Litter in submarine canyons off the west coast of Portugal. Deep Sea Research Part II: Topical Studies in Oceanography. 2011;**58**(23-24):2489-2496

[41] Aliani S, Molcard A. Hitch-hiking on floating marine debris: Macrobenthic species in the Western Mediterranean Sea. Hydrobiologia. 2003;**503**(1-3):59-67

[42] Barnes DKA, Fraser KPP. Rafting by five phyla on man-made flotsam in the Southern Ocean. Marine Ecology Progress Series. 2003;**262**:289-291

[43] Gregory MR. Environmental implications of plastic debris in marine settings—Entanglement, ingestion, smothering, hangers-on, hitch-hiking and alien invasions. Philosophical Transactions of the Royal Society B: Biological Sciences. 2009;**364**(1526):2013-2025

[44] Ritchie H, Samborska V, Roser M. Plastic Pollution. OurWorldInData.org. 2023 [cited 2024 Jul 19]. Available from: https://ourworldindata.org/plastic-pollution#article-citation

[45] Jambeck JR, Geyer R, Wilcox C, Siegler TR, Perryman M, Andrady A, et al. Plastic waste inputs from land into the ocean. Science (1979). 2015;**347**(6223):768-771. DOI: 10.1126/science.1260352

[46] Eriksen M, Lebreton LCM, Carson HS, Thiel M, Moore CJ, Borerro JC, et al. Plastic pollution

in the World's oceans: More than 5 trillion plastic pieces weighing over 250,000 tons afloat at sea. PLoS One. 2014;**9**(12):e111913. DOI: 10.1371/journal.pone.0111913

[47] Garcés-Ordóñez O, Castillo-Olaya VA, Granados-Briceño AF, Blandón García LM, Espinosa Díaz LF. Marine litter and microplastic pollution on mangrove soils of the Ciénaga Grande de Santa Marta, Colombian Caribbean. Marine Pollution Bulletin. 2019;**145**:455-462

[48] Martin C, Almahasheer H, Duarte CM. Mangrove forests as traps for marine litter. Environmental Pollution. 2019;**247**:499-508

[49] Zhang Y, Gao T, Kang S, Allen D, Wang Z, Luo X, et al. Cryosphere as a temporal sink and source of microplastics in the Arctic region. Geoscience Frontiers. 2023;**14**(4):101566. Available from: https://www.sciencedirect.com/science/article/pii/S1674987123000336

[50] Ramilo-Henry M, Umbelina B, Matilde E, Duncan EM. Plastic pollution on remote islands: A baseline study of Príncipe, gulf of Guinea. Marine Environmental Research. 2023;**191**:106181. Available from: https://www.sciencedirect.com/science/article/pii/S0141113623003094

[51] Barrett J, Chase Z, Zhang J, Holl MMB, Willis K, Williams A, et al. Microplastic pollution in Deep-Sea sediments from the great Australian bight. Frontiers in Marine Science. 2020:7. DOI: 10.3389/fmars.2020.576170

[52] Walther BA, Bergmann M. Plastic pollution of four understudied marine ecosystems: A review of mangroves, seagrass meadows, the Arctic Ocean and the deep seafloor. Emerging Topics in Life Sciences. 2022;**6**(4):371-387

[53] Crawford CB, Quinn B. Microplastic Pollutants. Amsterdam, Netherlands: Elsevier Limited; 2016

[54] Dike S, Apte S, Dabir V. Effect of low-density polyethylene, polyvinyl chloride, and high-density polyethylene micro-plastic contamination on the index and engineering properties of clayey soil-an experimental study. Environmental Research. 2023;**218**:115016. Available from: https://www.sciencedirect.com/science/article/pii/S001393512202343X

[55] Rincon-Rubio LM, Fayolle B, Audouin L, Verdu J. A general solution of the closed-loop kinetic scheme for the thermal oxidation of polypropylene. Polymer Degradation and Stability. 2001;**74**(1):177-188. Available from: https://www.sciencedirect.com/science/article/pii/S0141391001001549

[56] Lambert S, Wagner M. Characterisation of nanoplastics during the degradation of polystyrene. Chemosphere. 2016;**145**:265-268. Available from: https://www.sciencedirect.com/science/article/pii/S0045653515304094

[57] Adyel TM, Macreadie PI. World's largest mangrove Forest becoming plastic cesspit. Frontiers in Marine Science. 2021:8

[58] Jong MC, Tong X, Li J, Xu Z, Chng SHQ, He Y, et al. Microplastics in equatorial coasts: Pollution hotspots and spatiotemporal variations associated with tropical monsoons. Journal of Hazardous Materials. 2022;**424**:127626. Available from: https://www.sciencedirect.com/science/article/pii/S0304389421025942

[59] Gomez NCF, Cragg SM, Ghiglione JF, Onda DFL. Accumulation and exposure classifications of plastics in the different coastal habitats in the western Philippine archipelago. Environmental Pollution.

2023;**337**:122602. Available from: https://www.sciencedirect.com/science/article/pii/S0269749123016044

[60] Maquart PO, Froehlich Y, Boyer S. Plastic pollution and infectious diseases. Lancet Planet Health. 2022;**6**(10):e842-e845. Available from: https://www.sciencedirect.com/science/article/pii/S254251962200198X

[61] Tekman MB, Walther BA, Peter C, Gutow L, Bergmann M. Impacts of Plastic Pollution in the Oceans on Marine Species, Biodiversity and Ecosystems. Berlin, Germany: WWF; 2022. pp. 1-221

[62] Kannan G, Mghili B, Di Martino E, Sanchez-Vidal A, Figuerola B. Increasing risk of invasions by organisms on marine debris in the southeast coast of India. Marine Pollution Bulletin. 2023;**195**:115469. Available from: https://www.sciencedirect.com/science/article/pii/S0025326X23009037

[63] National Oceanic and Atmospheric Administration Marine Debris Program. Report on Marine Debris as a Potential Pathway for Invasive Species. National Oceanic and Atmospheric Administration Marine Debris Program. Silver Spring, MD; 2017

[64] Smith M, Love DC, Rochman CM, Neff RA. Microplastics in seafood and the implications for human health. Current Environmental Health Reports. 2018;**5**(3):375-386

[65] Parolini M, Stucchi M, Ambrosini R, Romano A. A global perspective on microplastic bioaccumulation in marine organisms. Ecological Indicators. 2023;**149**:110179

[66] Saikumar S, Mani R, Ganesan M, Dhinakarasamy I, Palanisami T, Gopal D. Trophic transfer and their impact of microplastics on estuarine food chain model. Journal of Hazardous Materials. 2024;**464**:132927

[67] Rochman CM, Tahir A, Williams SL, Baxa DV, Lam R, Miller JT, et al. Anthropogenic debris in seafood: Plastic debris and fibers from textiles in fish and bivalves sold for human consumption. Scientific Reports. 2015;**5**(1):14340

[68] Sharma S. Introductory chapter: Mangrove ecosystem research trends—Where has the focus been so far. In: Sharma S, editor. Mangrove Ecosystem Ecology and Function. London, UK: IntechOpen; 2018. Available from: https://www.intechopen.com/chapters/63980

[69] Giri C, Ochieng E, Tieszen LL, Zhu Z, Singh A, Loveland T, et al. Status and distribution of mangrove forests of the world using earth observation satellite data. Global Ecology and Biogeography. 2011;**20**(1):154-159

[70] Lee SY, Primavera JH, Dahdouh-Guebas F, McKee K, Bosire JO, Cannicci S, et al. Ecological role and services of tropical mangrove ecosystems: A reassessment. Global Ecology and Biogeography. 2014;**23**(7):726-743. Available from: https://www2.ulb.ac.be/sciences/biocomplexity/pub/Leeetal_2014_GEB.pdf

[71] Alongi DM, Murdiyarso D, Fourqurean JW, Kauffman JB, Hutahaean A, Crooks S, et al. Indonesia's blue carbon: A globally significant and vulnerable sink for seagrass and mangrove carbon. Wetlands Ecology and Management. 2015;**24**(1):3-13

[72] Azzarello MY, Van Vleet ES. Marine birds and plastic pollution. Marine Ecology Progress Series. 1987;**37**:295-303. Available from: https://www.int-res.com/articles/meps/37/m037p295.pdf

[73] Pruter AT. Sources, quantities and distribution of persistent plastics in the marine environment. Marine Pollution Bulletin. 1987;**18**(6):305-310

[74] Donohue MJ, Boland RC, Sramek CM, Antonelis GA. Derelict fishing gear in the northwestern Hawaiian islands: Diving surveys and debris removal in 1999 confirm threat to coral reef ecosystems. Marine Pollution Bulletin. 2001;**42**(12):1301-1312

[75] Moore CJ. Synthetic polymers in the marine environment: A rapidly increasing, long-term threat. Environmental Research. 2008;**108**(2):131-139

[76] van Sebille E, Aliani S, Law KL, Maximenko N, Alsina JM, Bagaev A, et al. The physical oceanography of the transport of floating marine debris. Environmental Research Letters. 2020;**15**(2):023003

[77] Diez SM, Patil PG, Morton J, Rodriguez DJ, Robin DV, et al. Marine Pollution in the Carbbean: Not a Minute to Waste. Washington, D.C: World Bank Group; 2019

[78] Monteiro RCP, Ivar do Sul JA, Costa MF. Plastic pollution in islands of the Atlantic Ocean. Environmental Pollution. 2018;**238**:103-110

[79] de Scisciolo T, Mijts EN, Becker T, Eppinga MB. Beach debris on Aruba, southern Caribbean: Attribution to local land-based and distal marine-based sources. Marine Pollution Bulletin. 2016;**106**(1-2):49-57

[80] Todd PA, Ong X, Chou LM. Impacts of pollution on marine life in Southeast Asia. Biodiversity and Conservation. 2010;**19**(4):1063-1082

[81] Green S, Webber M. A survey of the solid waste pollution in Kingston harbour mangroves, near port royal, Jamaica. Caribbean Marine Study. 1996;**5**:14-22

[82] Cordeiro CAMM, Costa TM. Evaluation of solid residues removed from a mangrove swamp in the São Vicente estuary, SP, Brazil. Marine Pollution Bulletin. 2010;**60**(10):1762-1767

[83] Costa MF, Silva-Cavalcanti JS, Barbosa CC, Portugal JL, Barletta M. Plastics buried in the inter-tidal plain of a tropical estuarine ecosystem. Journal of Coastal Research. 2011;**64**:339-343. Available from: http://www.jstor.org/stable/26482189

[84] Singare PU. Quantification study of non -biodegradable solid waste materials accumulated in the mangroves of Mahim Creek, Mumbai. Marine Sciences. 2012;**2**(1):1-5

[85] Debrot AO, Meesters HWG, Bron PS, de León R. Marine debris in mangroves and on the seabed: Largely-neglected litter problems. Marine Pollution Bulletin. 2013;**72**(1):1

[86] Vieira BP, Dias D, Nakamura EM, Arai TI, Hanazaki N. Is there temporal variation on solid waste stranding in mangroves? A case study in Ratones mangrove, Florianopolis, Brazil. Biotemas. 2013;**26**(1):79-86

[87] Hastuti AR, Yulianda F, Wardiatno Y. Spatial distribution of marine debris in mangrove ecosystem of Pantai Indah Kapuk, Jakarta. Bonorowo Wetlands. 2014;**4**(2):94-107

[88] Riascos JM, Valencia N, Peña EJ, Cantera JR. Inhabiting the technosphere: The encroachment of anthropogenic marine litter in Neotropical mangrove forests and its use as habitat by macrobenthic biota. Marine Pollution Bulletin. 2019;**142**:559-568

[89] Chee SY, Yee J, Danielle C, Yusup Y, Gallagher JB. Anthropogenic marine debris accumulation in mangroves on Penang Island, Malaysia. Journal of Sustainable Science and Management. 2020;**15**(6):36-60

[90] Hastuti AYUR, Lumbanbatu DTF, Wardiatno Y. The presence of microplastics in the digestive tract of commercial fishes off Pantai Indah Kapuk coast, Jakarta, Indonesia. Biodiversitas. 2019;**20**(5)

[91] Orth RJ, Carruthers TJB, Dennison WC, Duarte CM, Fourqurean JW, Heck KL, et al. A global crisis for seagrass ecosystems. Bioscience. 2006;**56**(12):987-996. DOI: 10.1641/0006-3568(2006)56[987:AGCFSE]2.0.CO

[92] Huang Y, Xiao X, Xu C, Perianen YD, Hu J, Holmer M. Seagrass beds acting as a trap of microplastics—Emerging hotspot in the coastal region? Environmental Pollution. 2020;**257**:113450. Available from: https://www.sciencedirect.com/science/article/pii/S0269749119316744

[93] Lamb JB, Willis BL, Fiorenza EA, Couch CS, Howard R, Rader DN, et al. Plastic waste associated with disease on coral reefs. Science (1979). 2018;**359**(6374):460-462. DOI: 10.1126/science.aar3320

[94] Macreadie PI, Nielsen DA, Kelleway JJ, Atwood TB, Seymour JR, Petrou K, et al. Can we manage coastal ecosystems to sequester more blue carbon? Frontiers in Ecology and the Environment. 2017;**15**(4):206-213. DOI: 10.1002/fee.1484

[95] Li C, Zhu L, Li WT, Li D. Microplastics in the seagrass ecosystems: A critical review. Science of the Total Environment. 2023;**902**:166152. Available from: https://www.sciencedirect.com/science/article/pii/S0048969723047770

[96] Gerstenbacher CM, Finzi AC, Rotjan RD, Novak AB. A review of microplastic impacts on seagrasses, epiphytes, and associated sediment communities. Environmental Pollution. 2022;**303**:119108. Available from: https://www.sciencedirect.com/science/article/pii/S0269749122003220

[97] Idris F, Febrianto T, Hidayati JR, Nugraha AH. Microplastic abundance in sea cucumber at seagrass ecosystem of Bintan Island and surrounding area, Indonesia. In: IOP Conference Series: Earth and Environmental Science. Bristol, England: IOP Publishing; 2022. p. 012009

[98] Nobre CR, Santana MFM, Maluf A, Cortez FS, Cesar A, Pereira CDS, et al. Assessment of microplastic toxicity to embryonic development of the sea urchin Lytechinus variegatus (Echinodermata: Echinoidea). Marine Pollution Bulletin. 2015;**92**(1-2):99-104

[99] Li P, Wang X, Su M, Zou X, Duan L, Zhang H. Characteristics of plastic pollution in the environment: A review. Bulletin of Environmental Contamination and Toxicology. 2021;**107**:577-584

[100] Ding L, Mao R f, Guo X, Yang X, Zhang Q, Yang C. Microplastics in surface waters and sediments of the Wei River, in the northwest of China. Science of the Total Environment. 2019;**667**:427-434. Available from: https://www.sciencedirect.com/science/article/pii/S004896971930840X

[101] Rodrigues JP, Duarte AC, Santos-Echeandía J, Rocha-Santos T. Significance of interactions between microplastics and POPs in the marine environment: A critical overview.

TrAC Trends in Analytical Chemistry. 2019;**111**:252-260. Available from: https://www.sciencedirect.com/science/article/pii/S0165993618305661

[102] Rochman CM, Browne MA, Halpern BS, Hentschel BT, Hoh E, Karapanagioti HK, et al. Classify plastic waste as hazardous. Nature. 2013;**494**(7436):169-171. Available from:. DOI: 10.1038/494169a

[103] Yoshida M. Social development and the environment—A view from solid waste management. In: International Development and the Environment: Social Consensus and Cooperative Measures for Sustainability. Singapore: Springer Singapore; 2020. pp. 27-43

[104] Praveenkumar TR, Sekar M, Pasupuleti RR, Gavurová B, Arun Kumar G, Vignesh KM. Current technologies for plastic waste treatment for energy recovery, it's effects on poly aromatic hydrocarbons emission and recycling strategies. Fuel. 2024;**357**:129379. Available from: https://www.sciencedirect.com/science/article/pii/S0016236123019932

[105] Hastings E, Potts T. Marine litter: Progress in developing an integrated policy approach in Scotland. Marine Policy. 2013;**42**:49-55. Available from: https://www.sciencedirect.com/science/article/pii/S0308597X13000419

[106] Liu TK, Chang H, Chen YS. Public awareness of marine environmental quality and its relationship for policy support on marine waste management. Marine Pollution Bulletin. 2023;**195**:115456. Available from: https://www.sciencedirect.com/science/article/pii/S0025326X23008901

[107] Katz D. Field Actions Science Reports Plastic Bank: Launching Social Plastic® revolution Creative Commons Attribution 3.0 License

[108] Ryan PG, Moore CJ, Van Franeker JA, Moloney CL. Monitoring the abundance of plastic debris in the marine environment. Philosophical Transactions of the Royal Society B: Biological Sciences. 2009;**364**(1526):1999-2012

Chapter 2

Restoring Maluku's Coast: Action-Driven Beach Cleanups, Mangrove Planting, and Coral Renewal in Ambon by the Faculty of Medicine, University of Pattimura

Bertha Jean Que, Parningotan Yosi Silalahi and Jerome Constantine Lekatompessy

Abstract

Maluku, Indonesia's archipelagic province, is predominantly maritime, with 92.4% ocean coverage. Its capital, Ambon, faces environmental challenges such as plastic waste accumulation, mangrove depletion, and coral reef degradation. Coastal areas, especially Ambon Bay, suffer from significant litter, predominantly plastic waste, leading to a "Very Dirty" classification by the Clean Coast Index. Mangrove forests, vital for shoreline protection and carbon sequestration, have declined by 2.98 hectares annually due to land conversion and human activity. Similarly, coral reefs have deteriorated over the past 35 years due to sedimentation and waste. The Faculty of Medicine at Universitas Pattimura plays a key role in environmental conservation efforts, including organized beach cleanups, mangrove planting, and coral reef restoration. These activities are integral to their curriculum, particularly through the Island Doctor program, which emphasizes marine and island health. Regular cleanups remove significant debris, while mangrove and coral restoration projects focus on long-term ecosystem rehabilitation. Despite challenges, these initiatives foster community awareness and demonstrate the potential for reversing environmental damage through coordinated action. This chapter highlights these conservation efforts and their impact on Maluku's coastal ecosystems.

Keywords: environmental challenges, beach cleanups, mangrove planting, coral restoration, island doctor program

1. Introduction

Maluku, an archipelagic province in Indonesia, is characterized by its predominantly maritime environment. Approximately 92.4% of Maluku's total area is ocean, with only 7.6% being land (**Figure 1**). Due to its vast number of islands and extensive

Figure 1.
The location of Maluku Province in the Indonesian archipelago [1].

sea coverage, Maluku is often referred to as the "Province of a Thousand Islands" nestled within the surrounding waters [2]. Ambon City serves as the capital of Maluku Province and is classified as one of Indonesia's island cities [3] with a significant portion of its area situated along the coastline of Ambon Bay [4]. Population growth in this capital could lead to higher resource demands, which often causes issues in land use. This expansion frequently harms the environment, neglecting sustainability and the ecosystem's carrying capacity [5]. In addition, this geographical setting faces significant environmental challenges, particularly concerning waste accumulation, mangrove depletion, and coral reef degradation.

Plastic was the most common type of litter found on the beaches. According to the Clean Coast Index (CCI), all coastal areas in Ambon Bay were classified as "Very Dirty," with significant amounts of waste covering the shorelines [6]. The majority of macroplastics found consisted of plastic film, primarily from food and beverage packaging and single-use plastic bags [7]. Coastal communities have taken steps to address waste issues, but their efforts have not fully resolved the problem [8]. Meanwhile, the coastline continues to struggle with challenges such as the loss of mangroves and the deterioration of coral reefs.

Mangrove forests are crucial coastal ecosystems that protect shorelines, prevent seawater intrusion, and absorb carbon [9]. Mangrove forests in Ambon Bay have been steadily shrinking by 2.98 hectares per year due to land conversion, other land use activities, significant abrasion risks, and ongoing human-induced disturbances [10]. Over the past decade, the biodiversity of the 20-hectare mangrove forest in Passo village, Ambon, has become increasingly concerning due to human-caused environmental damage [11]. Mangroves are at risk of degradation due to human activities such as overexploitation, waste dumping, and land conversion for economic development, all of which contribute to ecosystem decline [12].

Over the past 35 years, mangroves, seagrass, and coral reefs in Ambon Bay have shown significant declines in area, cover, and species diversity. Live coral coverage, which was once classified as "good" in 1986 (over 50% coverage), has diminished due to increased sedimentation and solid waste in the bay. Initially rated as "very good," live coral coverage has deteriorated to "good," then "amber," and in some areas, it has become critically low, with locations such as Halong and Kate Kate in Ambon Inner Bay showing 0% live coral coverage. This decline is primarily due to sedimentation from land clearing and solid waste [13].

To address the urgent need to reduce coastal waste and to protect and rejuvenate mangroves and coral reefs, immediate and effective action is essential. The Faculty of Medicine at the University of Pattimura, the sole medical faculty in Maluku Province, is pivotal in this effort. The Faculty of Medicine at the University of Pattimura has a vision to become a center of excellence for medical and health education, oriented toward island and marine competitiveness. It offers a distinctive course, known as the Island Doctor. This program not only aims to produce doctors who can help people but also equips them with the knowledge and skills necessary to work in marine and island areas, considering that Maluku Province is an archipelago.

Through organized beach cleanups, mangrove planting, and coral reef restoration, and in close collaboration with local nature and environmental groups, the faculty's students are making significant strides in safeguarding and restoring these crucial ecosystems. These activities are part of the curriculum, specifically the Island Doctor I course on anthropology, sociology, psychology, meteorology, and climatology, as well as the Island Doctor IV course on research into marine natural resources and the characteristics and health status of coastal and marine communities in the Maluku archipelago. This chapter delves into the specific actions undertaken by the Faculty of Medicine at the University of Pattimura, offering detailed insights into the initiatives implemented and the impacts achieved.

2. Beach cleanups

The problem of coastal debris has increasingly come into the spotlight in recent years, fueled by a combination of factors. These include poor waste management practices, a general lack of public awareness and education on the issue, insufficient large-scale initiatives, and weak enforcement of existing policies and regulations. Together, these elements contribute to the growing concern over the environmental and social impact of debris along coastlines [14]. Addressing this environmental challenge requires the immediate implementation of comprehensive strategies aimed at effectively managing and reducing the problem of coastal debris. It is essential to develop multifaceted approaches that not only target the cleanup of affected areas [15].

Beach cleanups by the Faculty of Medicine at the University of Pattimura students take place regularly, usually every two weeks or once a month, depending on the needs of the area. These efforts are often focused on villages that are supported by the Faculty of Medicine at the University of Pattimura. However, if other areas are found to require urgent attention, the clean-up teams are ready to mobilize to different beaches. The frequency of these clean-ups ensures a consistent approach to addressing waste accumulation, particularly in areas prone to heavy litter due to tides, wind, and seasonal factors.

A key focus during these clean-ups is the collection of waste in areas where trash tends to gather, especially along coastal regions near estuaries. These are particularly vulnerable because the trash discarded in urban areas often ends up in rivers, eventually making its way to the sea. From there, ocean currents transport this waste, and it frequently washes up on the beaches, particularly in bays where the water flow tends to deposit debris. This phenomenon has been observed repeatedly, making these coastal zones primary targets for clean-up efforts.

Before the beach cleanups, participants received proper preparation through orientation sessions. These sessions covered waste identification, safe collection methods, and handling environmental and health risks. Safety guidelines, including the use of protective gear and caution with hazardous materials, were emphasized. Volunteers were also instructed to record data on the waste collected, which would inform future waste management strategies. This preparation ensured that the clean-ups were effective and supported long-term environmental goals (**Figure 2**).

The amount of waste collected during each clean-up can vary significantly depending on the season, location, and recent weather patterns. However, it is common for these events to yield over 100 kilograms of trash. The types of waste found are diverse, though plastic products consistently make up the majority of the debris. Items such as food wrappers, detergent containers, and diapers are frequently collected, alongside larger objects such as discarded fishing nets, sacks, and even clothing. Household items, cans, and other metallic or glass objects also find their way into the piles of trash, further emphasizing the widespread issue of improper waste disposal. Earlier research has highlighted that Southeast Asian (SEA) nations rank among the leading contributors to the global production of solid municipal waste pollution. These countries generate approximately 1.14 kilograms of waste per person per day, contributing significantly to the worldwide waste problem [15].

Participation in these clean-ups tends to fluctuate, but an average of more than 20 people often join the efforts. Community involvement is a crucial part of the initiative, as it helps raise awareness about the impact of waste on the environment.

Figure 2.
Students participate in a cleanup, collecting debris and trash along a coastal area.

While these efforts have not yet been fully evaluated in terms of long-term results, there are already noticeable improvements in some areas. Several beaches that were previously littered with trash have become popular recreational spots for locals, a sign that the clean-ups are making a positive difference. Additionally, the visibility of the water in these areas has improved, which suggests that these ongoing efforts are helping to restore the natural beauty of the coastline.

These clean-ups are not just about removing trash but also about fostering a sense of responsibility within the community. With each event, more people are becoming aware of how their actions contribute to pollution and what can be done to prevent it. Though the road to lasting change is long, the visible improvements in water clarity and beach cleanliness offer hope for a healthier, more sustainable future for the coastal areas involved.

Multiple studies have emphasized the critical importance of community involvement in addressing the issue of marine debris. Notably, ongoing efforts across Southeast Asian countries aim to improve coastal cleanliness and reduce the presence of marine litter. A wide range of stakeholders—including governmental bodies, non-governmental organizations (NGOs), private sector entities, and local communities—are actively engaged in combating this problem. Their efforts include organizing beach clean-up activities, launching public education and awareness campaigns, and pushing for reforms in policies and regulations to strengthen waste management practices [16].

3. Mangrove planting

Mangroves represent a distinctive ecosystem, serving as a transition zone between land and sea, and are defined by extreme conditions such as high salinity levels, elevated temperatures, strong winds, powerful tides, and muddy, anaerobic soils. These ecosystems are both environmentally and economically crucial, particularly for many tropical and subtropical regions [17]. They offer significant social and ecological benefits to coastal communities, acting as natural barriers against storms and tsunamis, regulating local water systems, and providing essential habitats for a wide variety of marine species, including fish and other wildlife [18]. In addition to supporting biodiversity, mangroves are valuable sources of timber and non-timber forest products, contributing to both local economies and livelihoods. Furthermore, they offer nutritional resources and hold esthetic importance, making them attractive for ecotourism. Importantly, mangroves play a key role in combating climate change, as they possess a remarkable ability to store and sequester more carbon than terrestrial forests found in tropical and temperate zones [19].

The mangrove planting process adheres to established protocols that have been collaboratively developed by the Medical Assistance Team Vertebrae from the Faculty of Medicine at the University of Pattimura, along with nature enthusiast groups and the Environmental Agency. This cooperative effort ensures that the planting is carried out methodically and sustainably, incorporating best practices from environmental conservation while also leveraging the expertise and resources of each participating organization.

3.1 Location selection

The first step in the mangrove planting process is to identify an appropriate location for the seedlings. Generally, areas where mangroves already exist are prioritized, as they provide a supportive environment for new growth. However, careful coordination with nature enthusiast groups and the Environmental Agency is crucial to ensure that the chosen site aligns with broader conservation efforts and ecological goals. So far, three key locations—Lateri, Poka, and Kate-Kate—have been selected for mangrove planting. These sites were chosen based on their environmental conditions and suitability for fostering sustainable mangrove ecosystems. The collaboration between various stakeholders ensures that the restoration work is both effective and beneficial in the long term.

It is important to recognize that certain mangrove ecosystems experience high mortality rates due to various factors that impede their growth. One key reason for the failure of mangrove restoration efforts is the selection of unsuitable sites, where tidal inundation levels are insufficient to support healthy mangrove development. Additionally, strong wave activity can accelerate coastal erosion, leading to the death of newly planted mangroves. These environmental challenges highlight the need for careful site selection and consideration of local conditions in restoration projects (**Figure 3**) [20].

3.2 Seedling preparation

The next step involves preparing the mangrove seedlings, a process that is carried out with the support of the Environmental Agency. The mangrove seedlings to be planted are sourced from the Environmental Agency, where they are carefully selected and nurtured. These seedlings are chosen based on their suitability for the local environment. Not all seedlings are suited to every location, so it is important to assess and adjust the seedlings to match the specific conditions of the planting site. This preparation ensures that only the most appropriate seedlings are used, which enhances the chances of successful establishment.

Figure 3.
An area where mangroves already exist is prioritized as a planting site.

Figure 4.
Preparation of the planting site.

Once the seedlings are ready, the planting site itself must be prepared. This involves clearing the area to remove any debris or trash that might interfere with the planting process. By ensuring the site is clean and free from obstacles, the environment is better suited for the seedlings to take root and thrive. This thorough preparation helps to create optimal conditions for the mangroves, supporting their growth and contributing to the overall success of the reforestation efforts (**Figure 4**).

3.3 Seedling planting

A seedling from the Environmental Agency is classified by several key factors to ensure that it thrives in its new environment. These characteristics, such as species type, height, age, stage of development, and adaptation to the environment, play a crucial role in the seedling's survival and growth. First, the seedling is identified by its species. The Environmental Agency selects mangrove species that suit the local ecosystem, including common types such as *Rhizophora*, *Avicennia*, and *Sonneratia*. This decision is influenced by factors such as soil salinity, tidal patterns, and specific ecosystem needs.

The height of the seedling is another important factor in determining its readiness for planting. Mangrove seedlings between 30 and 50 cm are typically preferred. Seedlings that are too small may not be strong enough to survive, while those that are too tall may face difficulties adjusting due to underdeveloped root systems. Age is also critical, as it reflects the seedling's maturity and ability to cope with environmental stress. Seedlings are chosen when they are old enough to endure challenges such as changing water levels, but still young enough to adapt to a new location. Generally, this means they are several months to a year old, depending on the species.

The development stage of the seedling indicates whether it is ready for transplantation. Healthy roots, leaves, and stems are essential. Well-developed roots help the plant anchor in tidal zones, while vibrant leaves and stems signal that the seedling is more likely to survive and grow after being planted.

Finally, the seedling must be suited to the specific conditions of the planting site. The Environmental Agency evaluates factors such as salinity, water flow, and soil type to ensure that the seedlings are well-adapted to the local environment. Seedlings grown in similar conditions are much more likely to thrive in their new habitat.

By assessing these factors—species, height, age, development stage, and environmental suitability—the Environmental Agency ensures that the seedlings are strong and ready for successful coastal restoration projects.

Mangrove seedlings are typically planted in the morning when the tide is low. This timing is chosen to maximize the chances of successful planting, as lower tides make it easier to access the planting sites. The exact timing of these planting sessions is coordinated carefully with both nature enthusiast groups and the Environmental Agency. This collaboration ensures that the planting occurs under the most favorable conditions, taking into account factors such as tide schedules and environmental conditions.

To safeguard the young seedlings from potential threats such as debris or marine animals, it is recommended to use bamboo nets. These nets are placed around each seedling to provide a protective barrier that helps prevent damage and ensures that the seedlings have a better chance of establishing themselves in their new environment.

During a typical planting session, more than 1000 seedlings can be planted. However, this number is not fixed and is adjusted based on several factors. The specific conditions of the planting site play a crucial role in determining how many seedlings can be successfully planted. Additionally, the availability of seedlings and the amount of manpower available are also considered. The type of seedlings that can be used may vary depending on the location, as not all sites are suitable for the same types of seedlings. Furthermore, the timing of the tidal cycle influences how many seedlings can be planted in a given session (**Figure 5**).

For instance, in the planting activity carried out in Lateri, only 250 seedlings were used. This number was determined by the specific conditions at Lateri, including the site's suitability and the resources available at the time. This tailored approach ensures that each planting session is adapted to the unique characteristics of the location, thereby increasing the effectiveness of the mangrove restoration efforts.

Figure 5.
The Faculty of Medicine conducted mangrove seed planting during the COVID-19 pandemic (left) and after the pandemic (right).

3.4 Care and maintenance

Once the mangrove seedlings are planted, ongoing care and maintenance play a critical role in ensuring their survival and growth. Regular checks are needed to make sure that the bamboo nets used for protection are still intact and functioning as intended. It is important to inspect these nets for any damage and to clear away any debris that might have collected around the seedlings, as this can interfere with their development (**Figure 6**).

Keeping an eye on the seedlings for any signs of stress or damage is also vital. This involves monitoring their growth, looking out for any adverse effects from the environment, and addressing any issues as they arise. Maintaining good water quality and favorable tidal conditions around the planting site is essential for the health of the mangroves. If any seedlings show signs of struggling, additional care or even replanting might be necessary.

Engaging the local community in the care and maintenance of the mangrove area can significantly improve the success of the project. Involving residents in regular monitoring and upkeep helps ensure that the area remains clean and protected. Their participation fosters a sense of ownership and responsibility, which is crucial for the long-term success of the mangrove restoration efforts.

3.5 Monitoring and evaluation

After planting the mangrove seedlings, it is important to conduct an initial check after one to two weeks to assess their condition. This follow-up involves several key

Figure 6.
The bamboo net ensures the protection of the mangrove seed.

observations: Are the seedlings still intact? Is there any accumulated trash around them? Have they been damaged by insect bites? Additionally, evaluating the growth of the seedlings is crucial to understanding their development and ensuring they are established properly.

Regular monitoring is essential to address any issues promptly and support the seedlings' health. For example, in the Lateri planting site, out of 250 seedlings initially planted, it was reported that approximately 100 seedlings survived. This survival rate reflects the effectiveness of the initial planting efforts and the need for ongoing care.

The area successfully restored by these planting efforts can vary by location. While it is difficult to determine the exact size of the restored area for each site individually, combining all the locations gives an estimated total of around 1 hectare. This estimate provides a general idea of the scale of the restoration achieved, acknowledging that different sites have different conditions and challenges.

4. Coral restoration

4.1 Coral transplantation process

The first step in coral transplantation is selecting a suitable site for planting or transplantation. This involves studying the local ecosystem and current patterns to determine the most appropriate location. Once the site is chosen, a setup similar to an aquarium is prepared, taking into account the growth rate of different coral species.

Preparation for the transplantation involves setting up the planting environment on the beach. Coral fragments are then collected for transplantation. These fragments can be sourced from another area or nearby healthy coral patches, using a method that breaks the coral at its base to minimize damage. Once collected, the coral fragments are planted on the prepared setup, and arranged according to their growth rates (**Figure 7**). Fast-growing corals are placed at the bottom, while slower-growing corals are positioned toward the top. This arrangement is secured either with tying methods or special cement.

After the coral fragments are planted, their growth is monitored by measuring their length over time. Success is indicated by the increase in coral length, which reflects the effectiveness of the transplantation process.

4.2 Planning and setup

To date, there have been three areas where coral restoration efforts have been undertaken. Among these, two locations have shown promising results, with a survival rate of over 70%. However, one location has not yet shown significant improvement, primarily due to issues with current patterns and sediment that have hindered the growth of the coral polyps.

Ongoing efforts are focused on understanding these challenges better to improve the success rate of future projects. The team continues to study and refine techniques to enhance the effectiveness of coral restoration programs.

4.3 Restoration efforts and challenges

The planning and securing of coral fragments are carried out in shallow, submerged areas of the beach. This approach facilitates better arrangement and

Figure 7.
Preparation for the coral transplantation.

management of the coral fragments. Once all the coral pieces are positioned on the restoration media, this setup is promptly transferred to the designated planting site, following the planned layout (**Figure 8**).

The number of coral fragments used is generally based on the size of the area designated for restoration. The area's size is calculated according to the available budget, ensuring that resources are allocated effectively for each restoration project.

Figure 8.
The coral transplantation process involves attaching coral fragments to a metal frame.

Figure 9.
The My Coral event was attended by representatives from the Faculty of Medicine at the University of Pattimura.

In addition to the engagement of students and lecturers in coral restoration efforts at the beaches of Ambon, they also participate in similar activities conducted in various locations throughout Indonesia. One notable initiative is My Coral 5.0, which is part of the Community Service program at Management and Science University (MSU) in Malaysia. My Coral 5.0 has been an annual event organized by MSU for the past 5 years, aimed at advancing the United Nations' Sustainable Development Goals (SDGs). This initiative consists of four main components: community service, sea turtle conservation, coral planting, and mangrove restoration (**Figure 9**).

The activities involve students providing free health screenings to local residents and educating them about the importance of coral reefs. Participants also engage directly in the coral planting process. This work is vital, as many coral reefs, which serve as crucial habitats for fish, have been severely damaged due to human activities. Participants are organized into several groups, each receiving a metal frame that serves as a weight, along with coral seedlings to be securely attached to these frames using binding cables. Once all the corals are properly affixed, diving teams transport them to the ocean floor for installation.

On the same day as the coral planting, mangrove restoration activities are conducted. This aspect of the initiative is essential for maintaining healthy mangrove populations and preserving marine ecosystems, as mangroves play a critical role in coastal ecosystems. By engaging in both coral and mangrove restorations, the program aims to enhance biodiversity and contribute to the overall health of marine environments.

5. Conclusion

The coastal areas, shorelines, and marine ecosystems surrounding the city of Ambon are grappling with significant environmental challenges that necessitate urgent attention. Factors such as pollution, habitat degradation, and climate change are severely impacting these vital ecosystems, threatening both marine life and the livelihoods of local communities.

To effectively tackle the pressing issue of coastal waste and to protect and restore mangroves and coral reefs, immediate and decisive action is essential. The Faculty of Medicine at the University of Pattimura is making noteworthy strides in this area through a variety of well-organized initiatives. These include beach cleanups, which not only help to remove litter from the shorelines but also raise awareness about the importance of maintaining a clean coastal environment. Additionally, the faculty engages in mangrove planting and coral reef restoration projects, focusing on rehabilitating these critical habitats.

These efforts involve close collaboration with local environmental and conservation groups, ensuring a holistic approach to ecosystem preservation. By working together with community stakeholders, the initiatives are more likely to achieve sustainable results and foster a sense of shared responsibility for the environment.

Results from these activities indicate a marked improvement in beach cleanliness, with clearer water quality observed in the region. There has also been a notable increase in the number of mangrove seedlings planted, as well as a more extensive restoration of coral reefs. These positive outcomes underscore the effectiveness of coordinated efforts in revitalizing the coastal and marine environments of Ambon, paving the way for a healthier ecosystem that can support both wildlife and the communities that depend on these resources. Through continued commitment and action, there is hope for a more sustainable future for Ambon's coastal areas.

Acknowledgements

We extend our heartfelt gratitude to the students of the Faculty of Medicine at Universitas Pattimura, especially the Student Executive Board and the Medical Assistance Team Vertebrae, for their valuable contributions to this book chapter. Your dedication and insights were vital to its completion, and your commitment to environmental care initiatives is truly inspiring.

Author details

Bertha Jean Que*, Parningotan Yosi Silalahi and Jerome Constantine Lekatompessy
Faculty of Medicine University of Pattimura, Ambon, Indonesia

*Address all correspondence to: berthaque1966@gmail.com

References

[1] Perumahan & Kawasan Pemukiman Indonesia. PKP Maluku. 2020. Available at: https://perkim.id/profil-pkp/profil-perumahan-dan-kawasan-permukiman-provinsi-maluku/

[2] Kissiya E. Human's-planet connection in Maluku, Indonesia: Ecology anthropology perspective. In: Proceedings of the 3rd Borobudur International Symposium on Humanities and Social Science 2021 (BIS-HSS 2021). Paris: Atlantis Press SARL; 2023. pp. 426-430

[3] Nurhaeny A, Miharja M, Pradono DP. Measuring accessibility and island development in Ambon City. Island Studies Journal. 2021;**16**(1):373-387

[4] Setyawan WB, Barends W, Ainarwowan A, Polnaya D. Ambon City development trends: Its impact on the quality of the coastal environment of Ambon Bay and its vulnerability to hazards associated with sea-level rise. Pattimura Proceeding: Conference of Science and Technology. 2022;**2**:50-62

[5] Retraubun ASW, Selanno DAJ, Siahainenia L, Tubalawony S, Chr Tuhumury N, Sandhy DA. Coastal zone management of Passo Village of Ambon municipal, Indonesia. IOP Conference Series: Earth and Environmental Science. 2021;**805**(1):012020

[6] Yanti Manullang C, Barends W, Polnaya D, Soamole A, Rehalat I. Marine Litter and Grading of the Coastal Areas of Ambon Bay, Indonesia. 2021. Available at: http://dx.doi.org/10.20884/1.oa.2021.17.2.903

[7] Manullang CY, Soamole A, Rehalat I, Barends W, Sudin AM. Debris from plastic products in the beaches around northern Ambon Island. IOP Conference Series: Earth and Environmental Science. 2023;**1163**(1):012006

[8] Herdiansyah H, Saiya HG, Afkarina KII, Indra TL. Coastal community perspective, waste density, and spatial area toward sustainable waste management (case study: Ambon Bay, Indonesia). Sustainability. 2021;**13**(19):10947

[9] Alimudi S, Ohiwal M, Ainun NP, Kaliky B, Nurlette H, Ode WL. Detection of changes in Ambon Bay mangrove area using multitemporal satellite imagery. Jurnal Laot Ilmu Kelautan. 2023;**5**(2):193-203. Available from: www.jurnal.utu.ac.id/JLIK

[10] Kaimuddin MI, Kusmana C, Setiawan Y. Vegetation structure, biomass, and carbon of mangrove forests in Ambon Bay, Maluku, Indonesia. Jurnal Pengelolaan Sumberdaya Alam dan Lingkungan. 2023;**13**(4):710-722

[11] Latumahina FS, Susilawati S, Antonius A, Tewoloipaky W. Mangrove diversity as an indicator of ecosystem health on Ambon Island, Indonesia. Agro Bali: Agricultural Journal. 2024;**7**(2):295-306

[12] Sairmorsa W, Tetelepta EG. Identification of mangrove ecosystem damage in the PLTD Poka area. Ambon City. 2024;**3**(1):28-34

[13] Pelasula DD, Manullang CY, Patria MP, Wouthuyzen S, Lekallete JD, Malik SA. Destruction level on coral reef in the Ambon Bay. IOP Conference Series: Earth and Environmental Science. 2022;**1036**(1):012077

[14] Vince J, Hardesty BD. Governance solutions to the tragedy of the commons that marine plastics have become. Frontiers in Marine Science. 2018;**5**:5

[15] Ng CH, Mistoh MA, Teo SH, Galassi A, Ibrahim A, Sipaut CS, et al. Plastic waste and microplastic issues in Southeast Asia. Frontiers in Environmental Science. 2023;**11**:11

[16] Purba NP, Pasaribu B, Faizal I, Martasuganda MK, Ilmi MH, Febriani C, et al. Coastal clean-up in Southeast Asia: Lessons learned, challenges, and future strategies. Frontiers in Marine Science. 2023;**10**:10

[17] Njana MA. Structure, growth, and sustainability of mangrove forests of mainland Tanzania. Global Ecology and Conservation. 2020;**24**:e01394

[18] Wintah K, Hilmi E, MohH S. Mangrove diversity and its relationships with environmental conditions in Kuala Bubon Village, West Aceh, Indonesia. Biodiversitas. 2023;**24**(8):4599-4605

[19] Arifanti VB, Sidik F, Mulyanto B, Susilowati A, Wahyuni T, Subarno S, et al. Challenges and strategies for sustainable mangrove Management in Indonesia: A review. Forests. 2022;**13**(5):695

[20] Syahid LN, Sakti AD, Virtriana R, Wikantika K, Windupranata W, Tsuyuki S, et al. Determining optimal location for mangrove planting using remote sensing and climate model projection in Southeast Asia. Remote Sensing. 2020;**12**(22):3734

Chapter 3

Ecological Interactions and Adaptations of Polychaetes in Coastal Habitats of East Coast of India

Perumal Murugesan and Sasmita Swain

Abstract

Polychaetes are essential to marine ecosystems, acting as both prey and predators, sustaining food webs, and controlling invertebrate populations. Eunicidae and Nereididae are important predators in benthic habitats. Their digging and feeding activities promote bioturbation, which influences sediment structure, carbon oxidation, and nutrient cycling. Polychaetes have a variety of symbiotic partnerships, including with sea cucumbers and algae, which adds to ecological diversity. They are resistant to environmental stresses, with species such as *Capitella capitata* surviving in polluted areas, whilst others are more vulnerable to toxins. Adaptations such as bioturbation promote microbial diversity and assist in bioremediation. Human activities, including coastal expansion, pollution, and trawling, have a significant influence on the variety and quantity of polychaetes. Their ecological functions and flexibility emphasize their importance in sustaining healthy marine ecosystems, as well as the necessity for conservation measures to limit human effects.

Keywords: polychaetes, ecology, pollution indicator, bioturbation, Indian coast

1. Introduction

1.1 Background on polychaetes

Polychaetes, commonly referred to as bristle worms, are distinguished by the presence of numerous bristles, characteristic of the class Polychaeta of phylum Annelida. These organisms are characterized by their segmented bodies, with each segment possessing a pair of appendages called parapodia. Polychaetes are a diverse group of marine organisms exhibiting bilateral symmetry, various feeding strategies, and intricate sensory structures. Polychaetes have significant ecological roles and serve as indicators of environmental change.

Currently, as many as 12,677 living species have been recorded worldwide [1] with India comprising 727 species belonging to 334 genera and 72 families [2]. Polychaetes are broadly divided into two groups based on their adaptation, namely: Errantia (free moving) and Sedentaria (limited movement) [3]. Globally, the North Atlantic, Australia, and Indonesia are identified as the top three regions with the highest species richness of polychaetes. The southern hemisphere shows a higher total number of polychaete species compared to the northern hemisphere [4]. In India, polychaete diversity is reported to be high in the west coast compared to the east coast [5].

1.2 Why to study polychaetes?

Polychaetes hold significant importance in ecology and coastal ecosystems due to their impact on sediment processes and organic matter recycling [6]. These organisms are sensitive to environmental changes, and their presence, abundance, and diversity can serve as indicators of coastal health, pollution levels, and overall environmental conditions [7]. Polychaetes contribute to the stability of sediments, create habitats for other marine organisms, and support biodiversity by forming dense colonies [8]. Some polychaetes possess the ability to detoxify inorganic contaminants and process organically rich sediments, helping to maintain the health of coastal habitats [9].

In addition to their ecological roles, polychaetes are a vital component of the food web, with many organisms depending on them as a food source [10]. The simple body plan and regenerative abilities of polychaetes provide valuable insight into evolutionary developmental processes and regeneration mechanisms [11]. Compounds derived from polychaetes have also been proven to exhibit antimicrobial properties [12], and their high protein content makes them a valuable source of live feed in the aquaculture industry.

1.3 Data collection

This chapter synthesizes data from both published and gray literature sources to provide a comprehensive analysis of polychaete diversity and ecological interactions. The polychaete diversity data is specifically focused on the east coast of India, reflecting the unique species composition, habitat type and environmental conditions, of this region. In contrast, the data on ecological roles and interactions incorporates findings from a broader range of continental shelf regions and diverse marine habitats worldwide. This dual approach allows for a nuanced understanding of polychaetes within both local and global ecological contexts. In compiling this review, rigorous standards were maintained to ensure data quality and scientific integrity. Studies published in predatory journals were systematically excluded.

2. Diversity and distribution of polychaetes in East Indian coastal habitats

2.1 Overview of East Indian coastal regions

India's coastline, extending approximately 7515.6 km, hosts a wide variety of habitats characterized by significant biodiversity, in which the east coast of India contributes approximately 2545 km to the total Indian coastline [13].

The eastern coast, spanning from West Bengal in the northeast to Tamil Nadu in the southeast along the Bay of Bengal, is distinguished by its varied habitats, including expansive lush mangrove forests, serene lagoons, dynamic estuaries, intricate deltaic systems, and extensive sandy beaches [14], Specific regions like the Krishna-Godavari delta have a coastline of 525.15 km, featuring low-lying tidal mudflats, beaches, and mangrove swamps [15]. The Orissa coast, part of the east coast, has a coastline of 480 km, known for its estuaries, mangrove forests, and Chilika lake brackish water lagoons (**Figure 1**) [16].

2.2 Polychaete diversity

The diverse habitats stated above support a rich array of polychaete species, with variations in diversity and species richness observed across different coastal regions.

Numerous studies have investigated polychaete diversity in the deltaic and mangrove regions of the south coast of India. Misra et al. [17] conducted a pioneering study in the Gangetic delta, documenting 27 polychaete species across 18 genera and 12 families. In another study, Nandi and Choudhury [18] examined the benthic macrofauna of Sagar Island in the Sundarbans, identifying 17 species, with polychaetes being particularly well-represented. Common species included *Lumbrineris* sp., *L. polydesma*, *L. notocirrata*, *Diopatra neapolitana*, and *Talehsapia annandalei*. Misra and Choudhury [19] further explored the benthic fauna in specific Sundarbans mangrove locations, reporting 30 polychaete species from 24 genera and 13 families. Sarkar et al. [20] conducted research in two distinct areas within the Sundarbans Biosphere Reserve, Ghusighata and Gangasagar, documenting 11 species across six families, they identified unique polychaete assemblages: *Mastobranchus indicus* and *Dendronereides heteropoda* dominated sewage-influenced sediments at Ghusighata, while *Lumbrineris notocirrata*, *Ganganereis sootai*, and *Glycera tesselata* were prevalent at Gangasagar, near the Hooghly estuary's mouth, an area impacted by agricultural and industrial pollution, *Perinereis cultrifera* was primarily found during the post-monsoon period,

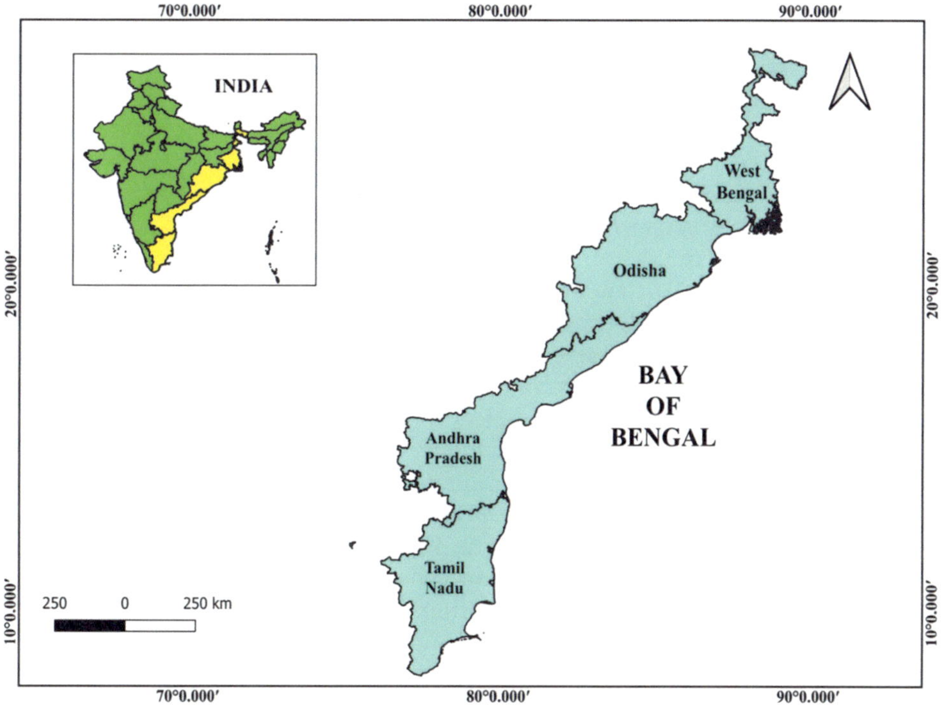

Figure 1.
Map showing the coastal habitats of East coast of India.

while *L. notocirrata* and *Diopatra cuprea* appeared occasionally. Rare or opportunistic species, including *Eteone barantollae*, *Namalycastes fauveli,* and *Ceratonereis burmensis*, emerged during the low-salinity monsoon period. Murugesan et al. [8] reviewed 68 polychaete species in the mangrove ecosystems along India's east coast, focusing significantly on the Sundarbans mangroves, where they recorded 39 species in the Muthupettai mangrove and 59 in the Pichavaram mangrove. Mitra et al. [21] concentrated on the intertidal macrofauna of the Subarnarekha estuary, documenting 25 polychaete species from 16 genera and 10 families, they noted that polychaetes primarily colonized both living and dead mangrove stems and roots, coexisting with other invertebrates. Nayak et al. [22] identified 17 polychaete species as the dominant group in the Atharbanki mangrove mudflats, with *Namalycastis* sp. notably abundant. Kathiresan [23] reported *Heteromastus similis*, *Euclymene annandale*, *Perinereis* sp., and *Mercierella enigmatica* in various Pichavaram mangrove habitats. Pravinkumar et al. [24] documented 16 polychaete species from the Pichavaram mangrove ecosystem. Murugesan et al. [25] compared benthic biodiversity between natural and artificially developed mangroves, recording 14 species in the Pichavaram mangrove and 11 in the artificially developed Vellar estuary mangroves. These studies collectively highlight the rich polychaete diversity and the influence of environmental factors on species distribution within India's mangrove ecosystems.

Few studies have focused on polychaete diversity in sandy habitats. Aiyar and Alikunhi [26] discovered a new species, *Pisionella indica*, on the sandy beaches of Chennai. Research on mud and mudflat habitats is similarly limited. Das et al. [27] documented 24 polychaete species from 14 genera and 12 families in the Digha-Talsari region of West Bengal. Many of these species were reported in the lower reaches, where substantial sewage and waste influxes enter the sea. *Lumbrineris polydesma* was found to dominate the muddy substrate of the upper littoral zone at Talsari, while *Parheteromastus* sp. was typically present in areas affected by sewage pollution.

Extensive research has been conducted on polychaete diversity in the estuarine waters along the eastern coast of India. Chakraborty and Choudhury [28] reported 14 polychaete species, spanning 13 genera and 8 families, from the Hooghly estuary at Sagar Island. Among these, *L. polydesma*, *G. tesselata*, *H. similis*, and *Parheteromastus tenuis* were found year-round, indicating their euryhaline nature and adaptability to varying salinity levels, in contrast, *Neanthes chingrighattensis* and *Capitellethus dispar* were abundant during the monsoon season, reflecting minimal tolerance for high salinity. Ajmalkhan and Murugesan [29] recorded 69 polychaete species in both the Hooghly-Matla and Mahanadi estuaries, establishing them as highly diverse Indian estuarine ecosystems for polychaetes. Roy et al. [30] documented six polychaete species from five genera and four families in the Matla Estuary. Siddique et al. [31] recorded 22 species from 20 genera and 15 families across various coastal and estuarine regions in Odisha and Andhra Pradesh. Rao et al. [32] documented 42 species from 32 genera and 16 families in the Vasishta Godavari estuary, with dominant species including *H. similis*, *Nephthys oligobranchia*, *Nereis lamellos*, *Indonereis gopalai*, *Dendronereis arborifera*, and *Magelona cincta*. Pandey et al. [33] assessed the ecological health of the Swaenamukhi River estuary, where polychaetes were the dominant group, comprising 41 taxa, with spionids being the most recurrent and abundant, constituting 28% of the macrofaunal community. Srikrishnadhas et al. [34] documented 15 polychaete species in the Vellar estuary and Parangipettai coastal waters. Chandran et al. [35] explored macrobenthos in the same area, identifying polychaetes as the dominant group, with 33 species observed, among these, *Nephthys polybranchia*, *Ancistrosyllis constricta*, *Lumbriconereis* sp., *Cossura delta*, *Glycera* sp., and *H. similis* were prevalent across all stations. Devi and Ayyakkannu [36] investigated the macrobenthos of the Buckingham Canal backwaters in the Coleroon estuary, identifying several polychaete species, including *H. similis*, *Ceratonereis costae*, *N. polybranchia*, *Glycera* sp., *Nereis* sp., *Goniada* sp., and *Capitella* sp. Jegadeesan and Ayyakkannu [37] researched polychaetes at two stations in the Coleroon estuary and the Bay of Bengal, with species like *N. polybranchia*, *Lumbrineris simplex*, *C. costae*, *H. similis*, and *Phyllodoce malmgranii* being dominant. Murugesan et al. [38] evaluated temporal changes in the benthic community of the marine zone in the Vellar estuary, where polychaetes, represented by 28 species, were the most abundant. In a subsequent study [39], they reported 13 species from a private shrimp farm on the southern bank of the Vellar estuary, including *A. constricta*, *N. polybranchia*, *Prionospio cirrifera*, and *C. delta*. Ramesh et al. [40] recorded 37 species in the Vellar estuary, with species such as *N. polybranchia*, *A. constricta*, and *C. delta* observed throughout the year. Muniasamy et al. [41] identified 44 species in the Coleroon estuary, where *Armandia* sp., *Cossura* sp., *Nephtys* sp., *Prionospio* sp., and *Syllis gracilis* were common across various stations. Balachandar et al. [42] explored polychaetes constituting 60% of the benthic community in Puducherry estuarine regions, with *Capitella capitata*, *Cossura coasta*, *Nephtys dibranchis*, and *Nereis capensis* being the dominant species. Selvaraj et al. [43] reported polychaetes as the predominant group, comprising 66% of the total benthic organisms in the Vellar and Ennore estuarine ecosystems. Sarathy et al. [44] recorded a total of 42 polychaete species, with the highest diversity observed in the Vellar estuary (41 species), followed by the Kaduvaiyar estuary (31 species), Ennore Creek (30 species), and the Uppanar estuary (25 species). These findings collectively underscore the rich polychaete diversity within the eastern estuarine ecosystems of India, influenced by both natural and anthropogenic factors, including pollution and seasonal salinity variations.

Several studies have been conducted on polychaete diversity in the coastal waters along the eastern coast of India. Chandra and Chakraborty [45] reported 18 polychaete species from 13 genera and 10 families along the Midnapore coast, finding the Errantia group to be more abundant than the Sedentaria group. Their study revealed that *T. annandalei*, *L. polydesma*, and *D. cuprea* reached peak densities during the pre-monsoon season, while *Perinereis nuntia*, *Eteone ornata*, and *M. indicus* were abundant during the monsoon, and *Glycera alba* and *Maldane sarsi* during the post-monsoon season. The study highlighted increased species dominance in the post-monsoon period, with peak diversity occurring in the pre-monsoon period. Balakrishnan and Tudu [46] compiled a checklist of polychaetes from the Odisha and West Bengal coasts, identifying 179 species across 8 orders, 36 families, and 119 genera, they observed that the order Phyllodocida included the highest number of families, while the family Nereididae contained the greatest species diversity. Hasan et al. [47] documented 40 species from 29 genera and 18 families along the Digha Coast. Numerous studies have also focused on polychaete diversity along the Odisha coast. Southern [48] conducted a pioneering benthic study in Chilika Lake, documenting 33 species from 25 genera and 13 families, while Soota and Rao [49] identified 25 species from 22 genera and 13 families along the Odisha coast. Ingole [50] reported 43 polychaete species from 20 genera in the coastal waters of Paradeep, with *Cossura longocirrata* emerging as the most abundant opportunistic species. Harkantra et al. [51] studied macrobenthos off the northeastern Bay of Bengal shelf, where polychaetes dominated, constituting 68.52% of the total macrobenthos. Common genera included *Eunice*, *Nephthys*, *Amphrodita*, *Glycera*, *Terebella*, *Sabellaria*, *Marphysa*, *Paranereis*, *Pseudonereis*, *Nereis*, *Owenia*, and *Diopatra*.

Srikrishnadhas [52] et al. recorded 123 species from 44 families and 94 genera in the Portonovo coastal waters, with 55 species classified as errantia and 68 as sedentaria. Similarly, Kundu et al. [53] found that polychaetes constituted 45% of the macrobenthic infaunal community along the Parangipettai coast. Musale and Desai [5] recorded 25 species along the south Indian coast, observing higher abundance on the southeast coast. Varadharajan et al. [54] identified 61 polychaete species dominating the benthic composition of the southeast coast. Anbuchezhian et al. [55] documented 34 species from the Thondi coastal waters, with the Glyceridae family most abundant near river mouths and the Eunicidae family dominating further offshore. Sivaleela and Venkataraman [56] recorded six genera and five families of polychaetes from 12 stations along the Tamil Nadu coast, with Capitellidae showing higher abundance. Manokaran et al. [57] reported 113 species of polychaetes from cruise collections made along the eastern coast of India. Recently, Balasubramanian et al. [58] documented 58 species from the Kanyakumari coast, with the Eunicidae and Nereidae families showing higher diversity in the region. These studies underscore the significant diversity of polychaete species along India's eastern coast, with variations in species composition and abundance influenced by season, habitat, and proximity to river mouths.

Coastal waters experience significant pollution, and polychaetes are widely recognized as indicators of such environmental stress. Notable studies include: Dipti et al. [59], who examined changes in benthic communities in response to pollution in Visakhapatnam's boat harbor, observed a decline in polychaete diversity—from 58 species in 1982–1984 to 23 species by 1994, in this period, species such as *C. coasta*, *tanaids*, and *Kalliapsuedes* were replaced by pollution-tolerant species like *Tharyx marioni* and *Nephtys oligobranchia*. In lagoon ecosystems, Nageswara Rao [60] documented 31 polychaete species from 24 genera and 14 families in Chilika Lake, a vital coastal lagoon. Similarly, Das [61] investigated how substratum types influence macrobenthic communities in Chilika's southern region, finding polychaetes to be the dominant group with 22 species, including *Nephthys* sp., *Glycera* sp., and *Capitella* sp. Mahapatro et al. [62] studied the effects of monsoonal rains on macrobenthic assemblages in Chilika Lagoon's outer channel, identifying polychaetes as the dominant taxa with eight prevalent species. These findings reinforce the role of polychaetes as key indicators of environmental health and pollution levels in coastal and lagoon ecosystems.

Coastal harbors serve as unique habitats for polychaetes, with several studies documenting their diversity in these environments. Sarma and Ganapati [63] recorded a total of 38 polychaete species from the northern side of Visakhapatnam Harbor. Expanding on this work, Raman and Ganapati [64] identified 72 species across 56 genera and 25 families within the same harbor, with prominent species including *C. capitata*, *Nereis glandicincta*, *C. coasta*, *T. marioni*, *Chloeia rosea*, *Axiothella australis*, *Chaetozone setosa*, and *N. dibranchis*. Subsequent studies in the region highlighted additional findings. Raman and Adiseshasai [65], along with Sivadas et al. [66], reported 50 polychaete species from the littoral areas near Visakhapatnam, with *D. neapolitana* emerging as a notable representative. Musale et al. [67] conducted an assessment of benthic macro-organisms in and around Visakhapatnam Harbor, confirming polychaetes as the dominant group with 60 species. Their study found that *Cirratulus* sp. was prevalent during the pre-monsoon season, while *C. coasta* and *N. dibranchis* were more abundant in the post-monsoon period. These studies underscore the considerable diversity of polychaetes within harbor environments and highlight seasonal shifts in species composition.

Several additional studies along the east coast of India have generated significant data on polychaete diversity and benthic ecosystems. Mitra and Misra [68] investigated the intertidal macrofauna at Talasari, documenting 10 polychaete species from 8 genera and 7 families. Pati et al. [69] compiled an extensive checklist based on 135 years of historical data for Odisha, listing 117 polychaete species from 114 genera and 43 families. Noyel and Desai [70] examined the macrobenthic population of Paradip Port, noting polychaetes as the most dominant group, comprising 70% of the total benthic community. They identified pollution indicator species such as *Tharyx* sp., *Prinospio* sp., *Cossura* sp., *Magelona* sp., and *Mediomastus* sp. Ganesh and Raman [71] studied the macrobenthic community of the northeast Indian shelf, where polychaetes represented 64.98% of the infauna, totaling 60 species. Manokaran et al. [72] recorded 75 species from 54 genera and 29 families in two areas, Cheyyur and Singarayakonda, with Spionids dominating in Cheyyur, and Ampharetids, Spionids, and Onuphids in Singarayakonda. Achari [73] described Micromaldane jonesi as a new species from the Gulf of Mannar. Further, Achari et al. [74] documented three families in the Gulf of Mannar and Palk Bay, including *Nothria mannarensis* from Pudumadam, *Scyphoproctus variabilis* from near Vedalai, and *Ancistargis brevicirris* from Palk Bay. In a study of the Palk Strait, Karthikeyan et al. [75] identified 38 polychaete species, with Syllidae, Ophellidae, and Eunicidae as the dominant families. Murugesan et al. [76] explored the utility of benthic diversity in assessing ecosystem health, finding that polychaetes constituted 30% of the total benthic communities in Tuticorin Bay and Vaan Island, Tamil Nadu. Samidurai et al. [77] reported 27 polychaete species, accounting for 58.69% of the macrobenthic population across three mangrove ecosystems in Tamil Nadu: developing, riverine, and island mangrove ecosystems. Prabakaran et al. [78] recorded 107 polychaete species across five stations along the Tamil Nadu coast. Most recently, Murugesan et al. [79] discovered a new species, *Pisione balasubramanian*, in Palk Bay, highlighting the continued significance of this region for polychaete research.

3. Ecological role and interactions

3.1 Role in food webs

Polychaetes, a highly diverse group of annelids, play pivotal roles in marine ecosystems, functioning as both prey and predators. As prey, they support a range of marine organisms, predominantly fin fish and mollusks. As predators, they help regulate populations of smaller invertebrates, contributing to the overall balance and health of benthic communities. In the tropical benthic carnivorous fish *Parupeneus barberinus*, polychaetes represent the most important prey item across all size classes, although they are generally selected against in favor of other prey such as ostracods, bivalves, and crabs [80]. Likewise, in the southern Bay of Biscay, polychaetes form a substantial component of the diet for 16 demersal fish species, with *Mullus surmuletus* standing out as a notable specialist on polychaetes, different fish species display distinct preferences for specific polychaete families and species, *M. surmuletus* predominantly preys on Nephtyidae, Sigalionidae, and Goniadidae families, while *Pagellus acarne* favors Sigalionidae, *Raja montagui* consumes polychaetes from the Sigalionidae, Arabellidae, Glyceridae, and Aphroditidae families, and *Scyliorhinus canicula* primarily targets polychaetes from the Arenicolidae, Nephtyidae, and Arabellidae families [81]. Numerous polychaete species, particularly those from the Eunicidae and Nereididae families, exhibit carnivorous behavior, they prey on smaller invertebrates, including mollusks and other polychaetes [82, 83]. The genus *Conus*, which consists of venomous marine gastropods, includes species that have adapted to specialize in feeding on polychaetes [84]. In the Maldives and Chagos Islands, polychaetes represent a primary food source for numerous *Conus* species, accounting for 96% of their diet [85]. Additionally, predatory gastropods, especially those in the Muricidae family, are known to consume polychaetes alongside other invertebrates [86]. Typhloscolecid polychaetes, such as *Typhloscolex muelleri* and *Travisiopsis dubia*, exhibit parasitic/predatory behavior by decapitating chaetognaths like *Flacisagitta enflata*, significantly impacting local populations [87]. Furthermore, the association between the sea urchin *Stomopneustes variolaris* and the polychaete *Lumbrineris latreilli* provides the polychaete protection from predators, and likely evolved from predation pressure [88]. In both the Andaman and Nicobar archipelago and the continental shelf of southeast India, predator-dominated polychaete assemblages are commonly found in well-oxygenated, coarse sandy sediments, particularly around mesophotic reefs at 50 m depth, where carnivorous polychaetes form a significant component of the macro benthic community [72, 89]. In ecosystems such as the Sundarbans, polychaetes contribute to the diets of species that are important for fisheries; this underscores their role in sustaining higher trophic levels in marine food webs [90].

3.2 Bioturbation and sediment dynamics

Polychaetes, a class of annelid worms, play a crucial role in shaping the structure and composition of marine sediments. Through their activities, such as bioturbation and feeding, they influence sediment characteristics and nutrient dynamics. Polychaete communities exhibit regional variations driven by factors like sediment texture, hydrodynamic conditions, and environmental stressors. For example, Sarkar et al. [20] identified distinct polychaete assemblages in two ecologically contrasting regions of the Sundarban Biosphere Reserve, which were influenced by differences in sediment composition and pollution levels. Key sediment characteristics such as grain size, organic content, and nutrient levels are significant determinants of polychaete distribution and community structure. Comparative studies by Musale and Desai [5] revealed that polychaete abundance is generally high along the west coast of India compared to the east coast. This difference is attributed to the loose texture of sediments in west coast, with high sand and sandy-silt content providing greater interstitial spaces for organisms to thrive. Thus, sediment type also greatly influences polychaete diversity: higher in coarse sandy sediments and lower diversity in fine, organic-rich sediments. Gopal et al. [89] observed that at a depth of 200 m, silty sediments with lower dissolved oxygen (DO) concentrations supported a higher abundance of deposit-feeding sedentary polychaetes, on the contrary, the mesophotic zones, characterized by well-oxygenated coralline sandy sediments, provided favorable microhabitats for hosting a variety of a number of predatory polychaetes in these coarse, well-oxygenated sediments, while deposit feeders dominated in deeper, low-oxygen environments.

Polychaetes significantly enhance metabolism in sediments and nutrient dynamics, especially in organically enriched sediments. Hansen and Kristensen [91] demonstrated that polychaete bioturbation increases oxygen penetration and promotes organic matter decomposition, leading to higher rates of carbon oxidation and nutrient cycling. Kristensen [92] further reported that the physical activities of polychaetes, such as *Nereis* spp. and *Arenicola marina*, greatly influence sediment carbon diagenesis. These activities enhance CO_2 flux in controlled laboratory settings, though the effect is less pronounced under natural conditions. Irrigation and sediment reworking by these polychaetes introduce oxygen into deeper sediment layers, stimulating carbon oxidation and reducing organic matter accumulation.

3.3 Symbiotic relationships

Polychaetes engage in various symbiotic relationships with other marine organisms, which can be mutualistic, commensal, or parasitic, and these interactions play important roles in marine ecosystems. For instance, *Axiothella mucosa* benefits from a symbiotic association with marine algae in its gelatinous egg masses. The algae provide food and oxygen, enhancing the fitness and survival of the developing embryos, while the algae utilize the eggs as a substrate for colonization and possibly as a nutrient source from decaying eggs [93]. Similarly, *Gastrolepidia clavigera* has been observed to form symbiotic relationships with holothurians (sea cucumbers) such as *Stichopus chloronotus*, *Holothuria atra*, and other species of Holothuria in the Lakshadweep Archipelago [94]. The solitary entoproct species, *Loxosomella diopatricola*, was recorded in the Bay of Bengal, off the southeastern coast of India, in a symbiotic relationship with the eunicid polychaete *Diopatra* sp. signaling a close association between the two species [95]. Polychaetes like *Terebella* sp., *Namalycastis abiuma*, and *Sabellaria* sp. are also known to host epibiotic ciliates, including *Epistylis* sp. and *Cothurnia* sp., on their integuments and parapodia, these associations further enhance our ken on the polychaete symbiosis and the complex interactions within marine ecosystems [96].

4. Response to environmental stressors

4.1 Pollution and contaminants

Polychaetes in Indian waters, much like those in other regions, are significantly affected by pollution and contaminants. As perturbation levels increase in a particular habitat, the sensitive species—having large body size with long life span, gets eliminated, while opportunistic species—characterized by smaller body sizes and shorter life spans—become dominant [97]. For example, the large bodied *Nereis virens* has been reported to accumulate organic contaminants in its tissues from polluted environments [98]. Some polychaete species are sensitive to high levels of organic matter, whereas others thrive in these enriched environments. Despite their adaptability, polychaetes remain highly vulnerable to pollution in Indian waters. True to this fact, the spionid polychaete, *Prionospio cirrobranchiata* was reported by Khan et al. [99] as an indicator species for industrial pollution. Oil spills and runoff containing hydrocarbons also pose significant threats to polychaetes, as they smother the organisms and contaminate their habitats. Ingole et al. [100] documented the negative impact of the grounded vessel MV River Princess oil spill on benthic community structures due to elevated petroleum concentrations at Goa beach sediments. However, bioturbation by polychaetes like *Hediste diversicolor* in oil-contaminated sediments can enhance microbial diversity and facilitate bioremediation [101].

A study in Thane Creek revealed that industrial, domestic, and waste pollution led to significant shifts in polychaete populations. Pollution-tolerant species such as *Lycastis ounaryensis* and *Polydora tentaculata* were totally replaced by more resilient/

conservative species like *C. burmensis*, driven by changes in sediment texture, low oxygen levels, and increased organic carbon and nitrogen concentrations [102]. Musale and Desai [5] also observed that certain species, such as *Prionospio pinnata* and *C. capitata*, which serve as indicators of organic pollution, exhibited higher abundance in areas with organic enrichment. This suggests that polychaetes can thrive in environments with elevated organic matter, often associated with pollution. For instance, in the mudflats of Shenzhen Bay, *Dendronereis pinnaticirris* and *Neanthes glandicincta* are negatively affected by high organic content, while species like *N. abiuma* and *C. capitata* thrive in such conditions [103]. Heavy metals, including mercury, lead, cadmium, and copper, are commonly found in Indian coastal and estuarine sediments and have been shown to harm marine organisms, including polychaetes. In the coastal waters of Nandgaon, Maharashtra, Sivaraj et al. [104] identified species such as *Boccardia polybranchia*, *C. capitata*, *Cirratulus cirratus*, *Heteromastus filiformis*, *Polydora ciliata*, *P. cirrifera*, and *P. pinnata* as being tolerant to toxic conditions. Similarly, in the Vellar-Coleroon estuarine system, which is impacted by various anthropogenic activities, species such as *C. capitata*, *Chaetozone* sp., *Cirratulus chrysoderma*, *Crassostrea madrasensis*, *Dorvillea rudolphi*, *Grandidierella* sp., *Mediomastus capensis*, and *Prionospio sexoculata* were identified as pollution-tolerant opportunistic species [105]. Different polychaete species show varying abilities to accumulate and tolerate heavy metals. For instance, *Branchiomma bairdi* accumulates high levels of heavy metals in its branchial crown, which may act as a defense mechanism against predators [106]. Similarly, *Nereis diversicolor* selectively accumulates metals linked to reduced sulfur compounds, such as pyrite, indicating a unique pathway for bioaccumulation [107]. The heavy metals like copper and mercury being particularly harmful to polychaetes such as *Neanthes arenaceodentata* and *C. capitata*, juvenile polychaetes are notably more sensitive to these metals compared to adults [108]. Pires et al. [109] found that *D. neapolitana*, a species often harvested as a natural resource, exhibited reduction in tissue regeneration and biochemical performance when exposed to contaminated sediment, with increased bioaccumulation leading to heightened lipid peroxidation (LPO) and lower glutathione (GSH/GSSG) levels.

Pollution also alters the community structure of polychaetes, especially in sensitive ecosystems like mangroves, where they play a crucial role in nutrient recycling and energy flow. Sediment contamination has been shown to negatively affect the physiological and biochemical performance of polychaetes. Penha-Lopes et al. [110] researched polychaete community in peri-urban equatorial and subtropical mangroves of East Africa, pollution in peri-urban mangroves leads to decreased diversity and shifts to more opportunistic polychaete species in Kenyan sites, while contaminated mangroves in Mozambique result in higher diversity.

In a yet another study by Selvaraj et al. [43], which assessed the ecological health of the Vellar and Ennore estuarine ecosystems using ecological health indices, several polychaete species were identified based on their tolerance to environmental conditions. The study found that *C. capitata*, *Cirratulus* sp., *Glycera unicornis*, *Lumbrineris magalhensis*, *Notomastus aberrans*, *P. cirrobranchiata*, and *P. pinnata* were tolerant species capable of surviving in moderately polluted environments. In contrast, species such as *Euclymene oerstedii*, *M. sarsi*, and *M. cincta* were highly sensitive to pollution. Additionally, dominant species such as *Eunice sciliensis*, *C. capitata*, *Cirratulus* sp., and *G. unicornis* demonstrated a strong association with heavy metal pollution and exhibited higher tolerance. Thus, pollution can disrupt the ecological processes mediated by polychaetes, affecting the overall health of ecosystems as has been stated by Murugesan and Balasubramanian [90]. Furthermore, the gametes of *Hydroides elegans* have been reported to be highly sensitive to environmental contaminants, making them useful for monitoring marine pollution in tropical and subtropical regions [111].

Microplastic pollution has also become rampant in the coastal waters of Tamil Nadu, particularly in Thoothukudi, Polychaetes, such as capitellids, have been found to ingest microplastics, leading to physical blockages and potential toxic effects from associated chemicals. High occurrences of polyethylene (PE) polymers, a common type of microplastic, were detected in polychaetes collected from Vellapatti and SPIC Nagar, Thoothukudi coastal waters [112].

Polychaetes have evolved diverse strategies to alleviate contaminants in their surroundings. They enhance their detoxification processes to cope with heavy metals and organic pollutants. In metal-contaminated environments, polychaetes can thrive by generating metallothioneins (MTs) or comparable metal-binding proteins [113]. These organisms can also process organic pollutants through their digestive systems and endure hypoxia, anoxia, and sulphidic conditions. A crucial adaptation is their capacity to dig deeper into the sediment, which shields them from surface pollutants and predators. This formation of biogenic structures not only safeguards them from harsh environmental conditions but also modifies the sediment's physical and chemical characteristics, establishing a more stable and conducive habitat [9]. In response to oxygen-depleted conditions resulting from nutrient pollution, polychaetes have developed morphological adaptations. Some species, particularly those from the Spionidae, Cossuridae, and Paraonidae families, display enlarged respiratory surfaces that allow them to survive in low-oxygen environments [114]. Polychaete species that are less tolerant to fluctuations in pCO_2 and pH are generally more vulnerable to environmental stressors. In a study by Sarathy et al. [115] on the effects of physico-chemical parameters and pCO_2 concentrations on mangrove-associated polychaetes in Pichavaram, it was found that elevated pCO_2 levels and low pH had a detrimental effect on polychaete diversity, resulting in reduced species richness. However, further analysis showed that *P. cirrifera*, *P. cirrobranchiata*, *P. sexoculata*, *Prionospio* sp., *C. capitata*, and *Orbiniid* sp. exhibited resilience to varying pCO_2 levels and were capable of surviving under these conditions. Although these species can tolerate elevated pCO_2 to some degree, their survival and overall fitness may be compromised under excessively high pCO_2 conditions.

4.2 Human activities and coastal development impact on polychaetes

Human activities and coastal development have significant impacts on polychaete populations. Discharges from Fish farms and effluents accrued from human settlements are reported to alter significantly polychaete abundance, richness, and diversity, with some families being more sensitive to pollution than others. Martínez et al. [116] observed that abundance, richness, and diversity of polychaetes significantly decrease under fish farm conditions, with polychaete distribution influenced by factors such as sulfide concentration, silt and clay content, and nitrogen isotope ratios, showing some families like Capitellidae, Dorvilleidae, Glyceridae, Nereididae, Oweniidae and Spionidae known to tolerate perturbation. In contrast, representatives from Magelonidae, Maldanidae, Nephtyidae, Onuphidae, Paralacydoniidae, Paraonide, Sabellidae, and Cirratulidae are sensitive to fish farm pollution. Urbanization and industrialization in Visakhapatnam Harbor have increased organic pollution, leading to distinct assemblages of polychaete benthic fauna with distance from the source: *C. capitata*, *N. glandicincta* in the inner harbor; *C. coasta*, *T. marioni* in the outer harbor; and a *Chloeia* sp., *Axiothella* sp., *Chaetozone* sp., and *Nephtys* sp. assemblage in the open sea [64].

In the coastal waters of Tamil Nadu, particularly in the Pichavaram and Palaverkadu regions, the Irular tribal fishing community intensively harvests polychaetes from their natural habitats which they sell to shrimp farms. This practice has led to concerns about resource depletion and the potential for overexploitation [117]. Muthuvelu et al. [118] studied the effects of bottom trawling on macrobenthic assemblages along the Cuddalore and Parangipettai coasts, finding a decline in the number of polychaete species before and after trawling—from 86 to 72 in Cuddalore and 76 to 62 in Parangipettai.

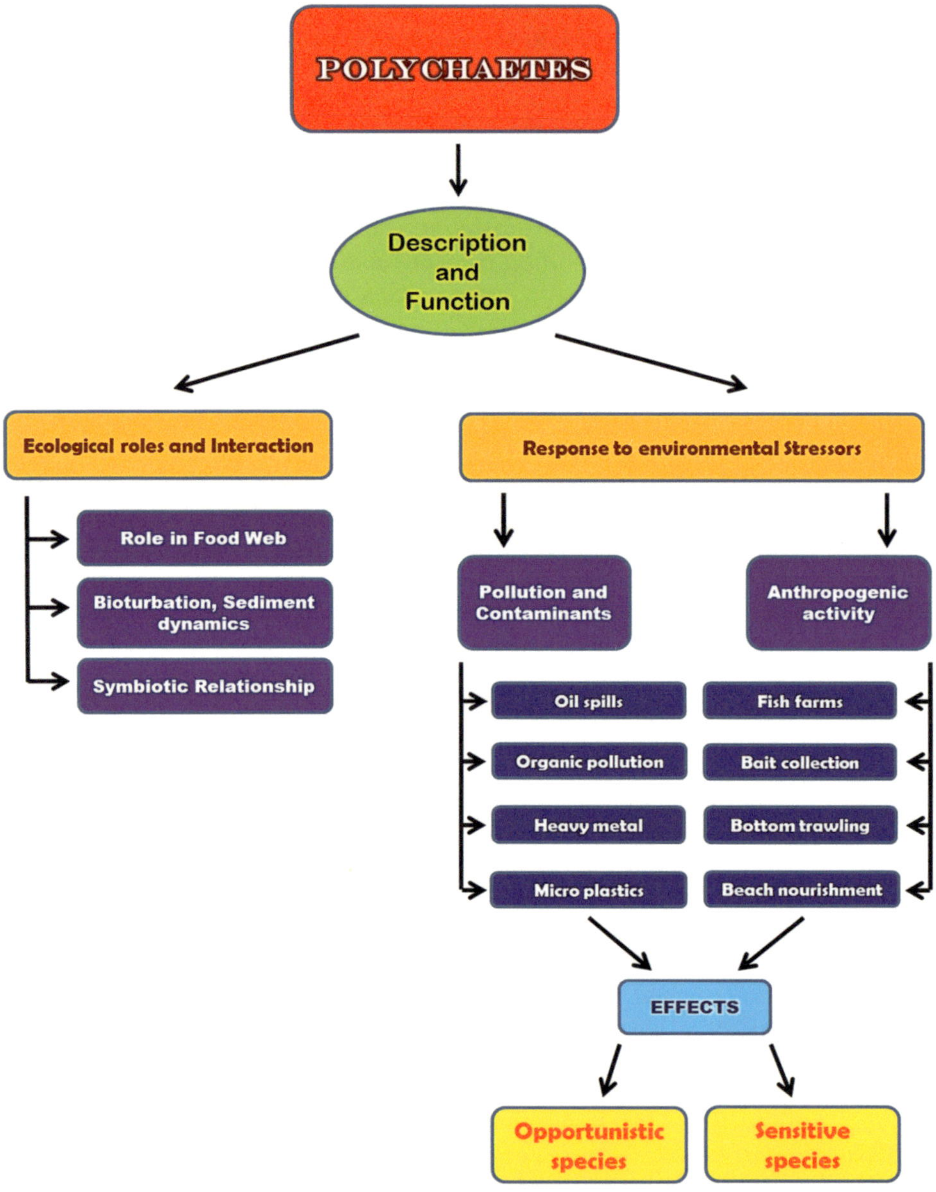

Figure 2.
Description and functional roles of polychaetes.

Beach nourishment, which involves installation of gryones with tripod or tetrapod structures to eroding coastlines, can also have telling effects on polychaete communities. Research by Danovaro et al. [119] suggests that small-scale beach replenishment is an environmentally sustainable method to combat coastal erosion, potentially affecting the abundance and structure of polychaete populations. The development of coastal areas often results in habitat fragmentation and degradation, leading to reduced biodiversity and declining polychaete numbers. These marine worms are vital to the ecosystem, serving as detritivores and food sources for other species. Therefore, with proper planning and suitable mitigative measures, the said beach replenishment could be initiated (**Figure 2**) [120].

5. Conclusion

Polychaetes in Indian waters exhibit remarkable resilience and adaptability to various environmental stressors, including pollution and contaminants. These marine organisms play vital roles in coastal ecosystems by contributing to nutrient cycling, sediment stability, and supporting biodiversity. However, the growing influence of anthropogenic interferences, such as industrial pollution, urbanization, and coastal development, has led to significant changes in polychaete community structures. The ability of certain species to thrive in organically enriched and polluted environments highlights their potential as bioindicators for monitoring ecosystem health. Adaptations such as enhanced detoxification mechanisms, burrowing behavior, and morphological changes enable polychaetes to survive in challenging conditions, yet continued exposure to pollutants, microplastics, and habitat degradation threatens their ecological roles. Effective management and conservation strategies need to be adopted to protect these species and maintain the balance and health of coastal ecosystems. By addressing pollution and preserving natural habitats, the polychaete diversity and the critical ecological functions they provide could be safeguarded.

Acknowledgements

We extend our sincere gratitude to R. Ragunna and S. Leo Lurison for their assistance in reference collection. We would also like to thank Anurag Sen for his valuable help in preparing and editing the figure for this manuscript.

Conflict of interest

The authors declare no conflict of interest.

Author details

Perumal Murugesan* and Sasmita Swain
Faculty of Marine Sciences, Centre of Advanced Study in Marine Biology, Annamalai University, Parangipettai, India

*Address all correspondence to: pmurugesaan74@gmail.com

References

[1] Read G, Fauchald K. World Polychaeta Database. 2024. Available from: https://www.marinespecies.org/polychaeta [Accessed: December 22, 2024]

[2] Sivadas SK, Carvalho R. Marine Annelida of India: Taxonomy and status evaluation and an updated checklist. Journal of Threatened Taxa. 2020;**12**(12):16647-16714. DOI: 10.11609/jot.5357.12.12.16647-16714

[3] Day JH. A Monograph on the Polychaeta of Southern Africa Part 1, Errantia: Part 2, Sedentaria: Trustees of the British Museum (Natural History). London: British Museum (Natural History); 1967. 878p

[4] Pamungkas J, Glasby CJ, Costello MJ. Biogeography of polychaete worms (Annelida) of the world. Marine Ecology Progress Series. 2019;**657**:147-159. DOI: 10.3354/meps13531

[5] Musale AS, Desai DV. Distribution and abundance of macrobenthic polychaetes along the South Indian coast. Environmental Monitoring and Assessment. 2011;**178**:423-436. DOI: 10.1007/s10661-010-1701-3

[6] Hutchings P. Biodiversity and functioning of polychaetes in benthic sediments. Biodiversity and Conservation. 1998;**7**:1133-1145

[7] Giangrande A, Licciano M, Musco L. Polychaetes as environmental indicators revisited. Marine Pollution Bulletin. 2005;**50**(11):1153-1162. DOI: 10.1016/j.marpolbul.2005.08.003

[8] Murugesan P, Sarathy PP, Muthuvelu S, Mahadevan G. Diversity and distribution of polychaetes in mangroves of east coast of India. In: Mangrove Ecosystem Ecology and Function. 6th chapter. London, UK: IntechOpen; 2018. pp. 107-130. DOI: 10.5772/intechopen.78332

[9] Fadhullah W, Syakir MI. Polychaetes as ecosystem engineers: Agents of sustainable technologies. In: Renewable Energy and Sustainable Technologies for Building and Environmental Applications: Options for a Greener Future. Springer International Publishing Switzerland: Springer Cham; 2016. pp. 137-150. DOI: 10.1007/978-3-319-31840-0_1

[10] Sobczyk R, Czortek P, Serigstad B, Pabis K. Modelling of polychaete functional diversity: Large marine ecosystem response to multiple natural factors and human impacts on the West African continental margin. Science of the Total Environment. 2021;**792**:148075. DOI: 10.1016/j.scitotenv.2021.148075

[11] Paulus T, Müller MC. Cell proliferation dynamics and morphological differentiation during regeneration in *Dorvillea bermudensis* (Polychaeta, Dorvilleidae). Journal of Morphology. 2006;**267**(4):393-403. DOI: 10.1002/jmor.10280

[12] Wu R, Patocka J, Nepovimova E, Oleksak P, Valis M, Wu W, et al. Marine invertebrate peptides: Antimicrobial peptides. Frontiers in Microbiology. 2021;**12**:785085. DOI: 10.3389/fmicb.2021.785085

[13] Muthusankar G, Lakshumanan C, Eswaramoorthi S. Inundation risk assessment for Nagapatinam Coast, Tamil Nadu, Southeast Coast of India. In: Monitoring and Prediction of Tropical Cyclones in the Indian

Ocean and Climate Change. Vol. 1. Dordrecht: Springer; 2014. pp. 364-373. DOI: 10.1007/978-94-007-7720-0_31

[14] Mukhopadhyay R, Karisiddaiah SM. The Indian coastline: Processes and landforms. In: Kale V, editor. Landscapes and Landforms of India. World Geomorphological Landscapes. Dordrecht: Springer; 2014. pp. 91-101. DOI: 10.1007/978-94-017-8029-2_8

[15] Pramanik MK, Biswas SS, Mondal B, Pal R. Coastal vulnerability assessment of the predicted sea level rise in the coastal zone of Krishna–Godavari delta region, Andhra Pradesh, east coast of India. Environment, Development and Sustainability. 2015;**18**:1635-1655. DOI: 10.1007/s10668-015-9708-0

[16] Mohanty PK, Panda US, Pal SR, Mishra P. Monitoring and management of environmental changes along the Orissa coast. Journal of Coastal Research. 2008;**24**:13-27. DOI: 10.2112/04-0255.1

[17] Misra A, Soota TD, Chowdhury A. On some polychaetes from Gangetic delta, West Bengal, India. Records of the Zoological Survey of India. 1983;**81**:41-54

[18] Nandi S, Choudhury A. Quantitative studies on the benthic macrofauna of Sagar Island, intertidal zones, Sunderbans, India. Mahasagar-Bulletin of the National Institute of Oceanography. 1983;**16**(3):409-414

[19] Misra A, Choudhury A. Polychaetous annelids from the mangrove swamps of Sunderbans, India. In: Bhosale LJ, editor. Proceedings on the National Symposium on Biological Utilization and Conservation of Mangroves. 1985. pp. 448-452

[20] Sarkar SK, Bhattacharya A, Giri S, Bhattacharya B, Sarkar D, Nayak DC, et al. Spatiotemporal variation in benthic polychaetes (Annelida) and relationships with environmental variables in a tropical estuary. Wetlands Ecology and Management. 2005;**13**:55-67. DOI: 10.1007/s11273-003-5067-y

[21] Mitra S, Misra A, Pattanayak JG. Intertidal macrofauna of Subarnarekha estuary (Balasore, Orissa). Zoological Survey of India. 2010;**313**:1-135

[22] Nayak A, Kumar BC, Lovaraju A, Raut D, Rout SS, Dash B, et al. Benthic infauna from mudflats of Atharbanki mangrove waterway in Odisha, India. Journal of the Marine Biological Association of India. 2018;**60**(1):34. DOI: 10.6024/jmbai.2018.60.1.2026-05

[23] Kathiresan K. A review of studies on Pichavaram mangrove, Southeast India. Hydrobiologia. 2000;**430**:185-205. DOI: 10.1023/A:1004085417093

[24] Pravinkumar M, Murugesan P, Prakash RK, Elumalai V, Viswanathan C, Raffi SM. Benthic biodiversity in the Pichavaram mangroves, Southeast Coast of India. Journal of Oceanography and Marine Science. 2013;**4**(1):1-1. DOI: 10.5897/JOMS12.004

[25] Murugesan P, Pravinkumar M, Muthuvelu S, Ravichandran S, Vijayalakshmi S, Balasubramanian T. Benthic biodiversity in natural vis-a-vis artificially developed mangroves of south east coast of India. Indian Journal of Geo-Marine Sciences. 2016;**45**(4):487-495

[26] Aiyar RG, Alikunhi KH. On a new pisionid from the sandy beach, Madras. Records of the Zoological Survey of India. 1940;**XLII**:89-107

[27] Das PA, Dhar PU, Chatterjee TK, Sarma DE. Species diversity of polychaete fauna of Digha-Talsari region of West Bengal, India. Journal

of the Inland Fisheries Society of India. 2009;**41**(1):16-20

[28] Chakraborty SK, Choudhury A. Occurrence and abundance of benthic polychaetes in Hooghly estuary, Sagar island, India. Journal of the Marine Biological Association of India. 1997;**39**:140-147

[29] AjmalKhan SA, Murugesan P. Polychaete diversity in Indian estuaries. Indian Journal of Marine Sciences. 2005;**34**(1):114-119

[30] Roy M, Nandi NC, Banerjee S, Majumder D. Distribution and abundance of macrozoobenthic species in some tropical Brackishwater wetlands of West Bengal, India. Proceedings of the Zoological Society. 2014;**67**:53-62. DOI: 10.1007/s12595-013-0071-6

[31] Siddique A, Purushothaman J, Bhowal A, Mandal S. Distribution and diversity of polychaeta (Phylum: Annelida) in the Northern coastal waters of Bay of Bengal. Records of the Zoological Survey of India. 2021;**121**(4):453-463. DOI: 10.26515/rzsi/v121/i4/2021/148098

[32] Rao DS, Rao MS, Annapurna C. Polychaete community structure of Vasishta Godavari estuary, east coast of India. Journal of the Marine Biological Association of India. 2009;**51**(2):137-144

[33] Pandey V, Venkatnarayanan S, Kumar PS, Ratnam K, Jha DK, Rajaguru S, et al. Assessment of ecological health of Swarnamukhi river estuary, southeast coast of India, through AMBI indices and multivariate tools. Marine Pollution Bulletin. 2021;**164**:112031. DOI: 10.1016/j.marpolbul.2021.112031

[34] Srikrishnadhas B, Jayabalan N, Ramamoorthi K. Ecology of the populations of polychaetes in the intertidal region of Vellar estuary. Proceedings Symposium Ecology Animal Popullation Zoological Survey of India. 1981;**1**:73-81

[35] Chandran R, Thangaraj GS, Sivakumar V, Srikrishnadhas B, Ramamoorthi K. Ecology of macrobenthos in the Vellar Estuary. Indian Journal of Marine Sciences. 1982;**11**:122-127

[36] Devi LP, Ayyakkannu K. Macrobenthos of the Buckingham Canal backwaters of Coleroon Estuary. Journal of the Marine Biological Association of India. 1989;**31**(1-2):80-85

[37] Jegadeesan P, Ayyakkannu K. Seasonal variation of benthic fauna in marine zone of Coleroon estuary and inshore waters, south east coast of India. Indian Journal of Marine Sciences. 1992;**21**:67-69

[38] Murugesan P, Ajmal Khan S, Ajithkumar TT. Temporal changes in the benthic community structure of the marine zone of Vellar estuary southeast coast of India. Journal of the Marine Biological Association of India. 2007;**49**:154-158

[39] Murugesan P, Ajithkumar TT, Khan SA, Balasubramanian T. Use of benthic biodiversity for assessing the impact of shrimp farming on environment. Journal of Environmental Biology. 2009;**30**(5):865

[40] Ramesh G, Murugesan P, Muthuvelu S. Research article diversity of benthic fauna in the marine zone of Vellar estuary (Southeast coast of India). International Journal Recent Scientific Research. 2013;**4**(12):2158-2166

[41] Muniasamy M, Muthuvelu S, Balachandar K, Murugesan P. Diversity

of benthic fauna in Coleroon estuary, south east coast of India. International Journal of Recent Scientific Research. 2013;**4**(10):1617-1621

[42] Balachandar K, Sundaramanickam A, Kumaresan S. Spatial and seasonal variation of macrobenthos from Puducherry Coast, Southeast Coast of India. International Journal of Current Microbiology and Applied Sciences. 2016;**5**:33-49. DOI: 10.20546/ijcmas.2016.510.006

[43] Selvaraj P, Murugesan P, Punniyamoorthy R, Parthasarathy P, Marigoudar SR. Assessment of the ecological health of Vellar and Ennore estuarine ecosystems using health indices. Indian Journal of Geo-Marine Sciences. 2019;**48**(10):1580-1592

[44] Sarathy PP, Bharathidasan V, Murugesan P, Selvaraj P, Punniyamoorthy R. Seasonal carbonate system vis-à-vis pH and salinity in selected tropical estuaries: Implications on polychaete diversity and composition towards predicting ecological health. Oceanologia. 2022;**64**(2):46-362. DOI: 10.1016/j.oceano.2022.01.001

[45] Chandra A, Chakraborty SK. Distribution, density and community ecology of macrobenthic intertidal polychaetes in the coastal tract of Midnapore, West Bengal, India. Journal of the Marine Biological Association of India. 2008;**50**(1):7-16

[46] Balakrishnan S, Tudu PC. An updated checklist of polychaetes (Animalia, Annelida, Polychaeta) from Odisha and West Bengal coasts. Indian Journal of Geo-Marine Sciences. 2021;**50**(06):507-510. DOI: 10.13140/RG.2.2.29972.42885

[47] Hasan N, Reynolds JW, Deuti K, Mandal CK, Misra A. Studies on the polychaetous annelids from Digha coast, West Bengal, India. Megadrilogica. 2022;**27**(5):42-56

[48] Southern R. Fauna of the Chilka Lake and also of fresh and brackish waters in other parts of India. Memoirs of the Indian Museum Calcutta. 1921;**5**:563-659

[49] Soota TD, Rao CN. On some polychaetes from Orissa coast. Records of the Zoological Survey of India. 1977;**74**(4):327-336

[50] Ingole BS. Biodiversity of benthic polychaetes from the coastal waters of Paradip, Bay of Bengal. Bionano Frontier. 2007;**1**(1):32-45

[51] Harkantra SN, Rodrigues CL, Parulekar AH. Macrobenthos of the shelf off North Eastern Bay of Bengal. Indian Journal of Marine Sciences. 1982;**11**:115-121

[52] Srikrishnadhas B, Ramamoorthi K, Balasubrahmanyan K. Polychaetes of Porto Novo waters. Journal of the Marine Biological Association of India. 1987;**29**:134-139

[53] Kundu S, Mondal N, Lyla PS, Ajmal KS. Biodiversity and seasonal variation of macro-benthic infaunal community in the inshore waters of Parangipettai Coast. Environmental Monitoring and Assessment. 2010;**163**:67-79. DOI: 10.1007/s10661-009-0817-9

[54] Varadharajan D, Soundarapandian P, Gunalan B, Babu R. Seasonal abundance of macro benthic composition and diversity along the south east coast of India. European Journal of Applied Sciences. 2010;**2**(1):1-5

[55] Anbuchezhian RM, Ravichandran S, Murugesan P, Serebiah JS. Assessment

of soft bottom polychaete diversity in Thondi, Palk Bay, India. Journal of Environmental Biology. 2012;**33**(5):917-921

[56] Sivaleela G, Venkataraman K. Diversity and distribution of polychaetes from Tamil Nadu Coast, India. In: Ecology and Conservation of Tropical Marine Faunal Communities. Berlin, Heidelberg: Springer; 2013. pp. 55-62. DOI: 10.1007/978-3-642-38200-0_4

[57] Manokaran S, Khan SA, Lyla S, Raja S, Ansari KG. Feeding guild composition of shelf macrobenthic polychaetes of southeast coast of India. Tropical Zoology. 2013;**26**(3):120-139. DOI: 10.1080/03946975.2013.825425

[58] Balasubramanian S, Soniya S, Ahamed SJ, Babu MM, Rajan RA, Citarasu T. Taxonomic list of macro benthic polychaetes of Kanyakumari Coast. Uttar Pradesh Journal of Zoology. 2023;**44**(23):416-429. DOI: 10.56557/UPJOZ/2023/v44i233802

[59] Dipti R, Rao PE, Raman AV. Assessment of benthic community alterations in relation to pollution in a boat harbor in Visakhapatnam, East Coast of India. The Ecoscan. 2013;**7**(1&2):51-56

[60] Nageswara Rao CA. Annelida: Polychaeta. In: Ghosh AK, editor. Fauna of Chilka Lake. Zoological Survey of India. Wetland Ecosystem Series 1. Calcutta, India: Zoological Survey of India; 1995. pp. 319-336

[61] Das S. Effect of substratum on the community structure of macrobenthic fauna from southern sector of Chilika Lake. Indian Journal of Fisheries. 2004;**51**(3):375-380

[62] Mahapatro D, Panigrahy RC, Panda S, Mishra RK. Influence of monsoon on macrobenthic assemblage in outer channel area of Chilika lagoon, east coast of India. Journal of Wetlands Ecology. 2009;**3**:56-67. DOI: 10.3126/jowe.v3i0.2223

[63] Sarma ALN, Ganapati PN. Meiofauna of the Visakhapatnam harbour in relation to pollution. Bulletin of the Department of Marine Science, University of Cochin. 1975;**7**:243-255

[64] Raman AV, Ganapati PN. Pollution effects on ecobiology of benthic polychaetes in Visakhapatnam Harbour (Bay of Bengal). Marine Pollution Bulletin. 1983;**14**(2):46-52. DOI: 10.1016/0025-326X(83)90190-X

[65] Raman AV, Adiseshasai K. Macrobenthos from littoral areas off Visakhapatnam, east coast of India. Indian Journal of Marine Sciences. 1989;**18**(4):265-269

[66] Sivadas SK, Ingole BS, Fernandes CE. Environmental gradient favours functionally diverse macrobenthic community in a placer rich tropical bay. The Scientific World Journal. 2013;**2013**:750580. DOI: 10.1155/2013/750580

[67] Musale AS, Desai DV, Sawant SS, Venkat K, Anil AC. Distribution and abundance of benthic macroorganisms in and around Visakhapatnam Harbour on the east coast of India. Journal of the Marine Biological Association of the United Kingdom. 2015;**95**(2):215-231. DOI: 10.1017/S0025315414001490

[68] Mitra S, Misra A. A report on intertidal macrofauna of Talsari (Balasore: Orissa). Records of the Zoological Survey of India. 2006;**106**(3):131-141. DOI: 10.26515/rzsi/v106/i3/2006/159194

[69] Pati SK, Swain D, Sahu KC, Sharma RM, Mohapatra A. Marine

fauna of Odisha, east coast of India: An annotated checklist of historical data of 135 years. Journal of Aquatic Biology and Fisheries. 2018;**6**:1-15

[70] Noyel V, Desai DV. Spatio-temporal variation in the macrobenthos of Paradip port, east coast of India. Current Science. 2020;**119**(1):68-82. DOI: 10.18520/cs/v119/i1/68-82

[71] Ganesh T, Raman AV. Macrobenthic community structure of the northeast Indian shelf, Bay of Bengal. Marine Ecology Progress Series. 2007;**341**:59-73. DOI: 10.3354/meps341059

[72] Manokaran S, Khan SA, Lyla PS. Macrobenthic composition of the southeast continental shelf of India. Marine Ecology. 2014;**36**(1):1-15. DOI: 10.1111/maec.12107

[73] Achari GPK. Studies on new or little known Polychaetes from the Indian seas. 2. Micromaldane jonesi n. Sp. (Maldanidae). Journal of the Marine Biological Association of India. 1968;**10**(2):269-273

[74] Achari GPK. Catalogue of polychaetes in the reference collections of the Central Marine Fisheries Research Institute. In: James PSBR, Thomas PA, Pillai CSG, Achari GPK, editors. Catalogue of Types and of Sponges, Corals, Polychaetes, Crabs and Echinoderms in the Reference Collections of the Central Marine Fisheries Research Institute. Bulletin of CMFRI. Vol. 7. Mandapam, India: Central Marine Fisheries Research Institute; 1969. pp. 31-40

[75] Karthikeyan MM, Prakash RK, Ananthan G. Macro benthic assemblage and temporal interactions at Palk Strait, southeast coast of India. World Journal of Zoology. 2009;**4**(2):96-104

[76] Murugesan P, Muniasamy M, Muthuvelu S, Vijayalakshmi S, Balasubramanian T. Utility of benthic diversity in assessing the health of an ecosystem. Indian Journal of Geo-Marine Sciences. 2011;**40**(3):243-249

[77] Samidurai K, Saravanakumar A, Kathiresan K. Spatial and temporal distribution of macrobenthos in different mangrove ecosystems of Tamil Nadu Coast, India. Environmental Monitoring and Assessment. 2012;**184**:4079-4096. DOI: 10.1007/s10661-011-2245-x

[78] Prabakaran JR, Vibin A, Deivakumari M, Muruganantham M, Ramasubburayan R, Palavesam A, et al. Comparison of polychaete diversity and distribution along the south Tamilnadu coast (Lat. 8.08° to 10.79° N), India. Regional Studies in Marine Science. 2019;**28**:100564. DOI: 10.1016/j.rsma.2019.100564

[79] Murugesan P, Miranda R, Punniyamoorthy R, Mahadevan G. A new species of Pisione (Annelida, Sigalionidae) from Palk Bay, India. Journal of the Marine Biological Association of the United Kingdom. 2023;**103**(55):1-9

[80] Lukoschek V, McCormick M. Ontogeny of diet changes in a tropical benthic carnivorous fish, Parupeneus barberinus (Mullidae): Relationship between foraging behaviour, habitat use, jaw size, and prey selection. Marine Biology. 2001;**138**:1099-1113. DOI: 10.1007/S002270000530

[81] Serrano A, Velasco F, Olaso I. Polychaete annelids in the diet of demersal fish from the southern shelf of the Bay of Biscay. Journal of the Marine Biological Association of the United Kingdom. 2003;**83**:619-623. DOI: 10.1017/S0025315403007550h

[82] Clemo W, Dorgan K. Functional morphology of eunicidan (polychaeta)

jaws. The Biological Bulletin. 2016;**233**:227-241. DOI: 10.1086/696291

[83] Pardo E, Dauer D. Particle size selection in individuals from epifaunal versus infaunal populations of the nereidid polychaete Neanthes succinea (Polychaeta: Nereididae). Hydrobiologia. 2003;**496**:355-360. DOI: 10.1007/978-94-017-0655-1_34

[84] Duda J, Kohn A, Palumbi S. Origins of diverse feeding ecologies within Conus, a genus of venomous marine gastropods. Biological Journal of the Linnean Society. 2001;**73**:391-409. DOI: 10.1111/J.1095-8312.2001.TB01369.X

[85] Khon AJ. Microhabitats, abundance and food of Conus on atoll reefs in the Maldive and Chagos Islands. Ecology. 1968;**49**(6):1046-1062. DOI: 10.2307/1934489

[86] Taylor J. Habitats and diet of predatory gastropods at addu atoll, Maldives. Journal of Experimental Marine Biology and Ecology. 1978;**31**:83-103. DOI: 10.1016/0022-0981(78)90138-7

[87] Øresland V, Bray RA. Parasites and headless chaetognaths in the Indian Ocean. Marine Biology. 2005;**147**(3):725-734. DOI: 10.1007/s00227-005-1618-5

[88] Ayyagari A, Kondamudi RB. Ecological significance of the association between *Stomopneustes variolaris* (Echinoidea) and *Lumbrineris latreilli* (Polychaeta) from Visakhapatnam coast, India. Journal of Marine Biology. 2014;**2014**:1-4. DOI: 10.1155/2014/640785

[89] Gopal A, Abdul Jaleel KU, Parameswaran UV, Sanjeevan VN, Saramma AV, Vijayan A, et al. Distinctive community patterns with exceptional diversity of polychaetes around a tectonically active archipelago in the tropical Indian Ocean. Frontiers in Marine Science. 2020;**7**:710. DOI: 10.3389/fmars.2020.00710

[90] Murugesan P, Balasubramanian T. Diversity and community structure of polychaetes in mangroves of Indian Coast. In: Mangroves: Ecology, Biodiversity and Management. Vol. 14. Singapore: Springer; 2021. pp. 323-359. DOI: 10.1007/978-981-16-2494-0_14

[91] Hansen K, Kristensen E. The impact of the polychaete Nereis diversicolor and enrichment with macroalgal (*Chaetomorpha linum*) detritus on benthic metabolism and nutrient dynamics in organic-poor and organic-rich sediment. Journal of Experimental Marine Biology and Ecology. 1998;**231**:201-223

[92] Kristensen E. Impact of polychaetes (Nereis spp. and Arenicola marina) on carbon biogeochemistry in coastal marine sediments. Presented during the ACS Division of Geochemistry symposium 'Biogeochemical Consequences of Dynamic Interactions Between Benthic Fauna, Microbes and Aquatic Sediments', San Diego, April 2001. Geochemical Transactions. 2001;**2**(12):92-103. DOI: 10.1039/b108114d

[93] Peyton KA, Hanisak MD, Lin J. Marine algal symbionts benefit benthic invertebrate embryos deposited in gelatinous egg masses. Journal of Experimental Marine Biology and Ecology. 2004;**307**(2):139-164. DOI: 10.1016/j.jembe.2004.01.022

[94] Marudhupandi T, Kumar TT, Prakash S, Gopi M, Balasubramanian T. A first report of symbiotic polychaete Scale Worm Gastrolepidia clavier Schmarda, 1861 (Phyllodocida: Polynoidae) from Lakshadweep Archipelago, India. Journal of

Threatened Taxa. 2014;**6**(10):6385-6388. DOI: 10.11609/JoTT.o3092.6385-8

[95] Bharathidasan V, Borisanova AO, Sarathy PP, Murugesan P. New entoproct-polychaete association recorded in the Bay of Bengal, southeastern coast of India. Zootaxa. 2022;**5100**(4):482-500

[96] Bharathidasan V, Sarathy PP, Murugesan P, Nogueira JM. New records for associations between peritrich protozoan ciliates (Ciliophora, Sessilida) and polychaete worms (Annelida) from off the southeastern coast of India. Zootaxa. 2021;**5071**(4):492-504. DOI: 10.11646/zootaxa.5071.4.2

[97] Warwick R. A new method for detecting pollution effects on marine macrobenthic communities. Marine Biology. 1986;**92**:557-562. DOI: 10.1007/BF00392515

[98] Pruell RJ, Taplin BK, McGovern DG, McKinney R, Norton SB. Organic contaminant distributions in sediments, polychaetes (*Nereis virens*) and American lobster (*Homarus americanus*) from a laboratory food chain experiment. Marine Environmental Research. 2000;**49**(1):19-36. DOI: 10.1016/S0141-1136(99)00046-X

[99] Khan SA, Murugesan P, Lyla PS, Jaganathan S. A new indicator macro invertebrate of pollution and utility of graphical tools and diversity indices in pollution monitoring studies. Current Science. 2004;**87**(11):1508-1510

[100] Ingole B, Sivadas S, Goltekar R, Clemente S, Nanajkar M, Sawant R, et al. Ecotoxicological effect of grounded MV River Princess on the intertidal benthic organisms off Goa. Environment International. 2006;**32**(2):284-291

[101] Taylor JD, Cunliffe M. Polychaete burrows harbour distinct microbial communities in oil-contaminated coastal sediments. Environmental Microbiology Reports. 2015;7(4):606-613. DOI: 10.1111/1758-2229.12292

[102] Quadros G, Sukumaran S, Athalye RP. Impact of the changing ecology on intertidal polychaetes in an anthropogenically stressed tropical creek, India. Aquatic Ecology. 2009;**43**(4):977-985. DOI: 10.1007/s10452-009-9229-8

[103] Cai LZ, Hwang JS, Dahms HU, Fu SJ, Chen XW, Wu C. Does high organic matter content affect polychaete assemblages in a Shenzhen Bay mudflat, China? Journal of Marine Science and Technology. 2013;**21**:274-284. DOI: 10.6119/JMST-013-1223-5

[104] Sivaraj S, Murugesan P, Muthuvelu S, Vivekanandan KE, Vijayalakshmi S. AMBI and M-AMBI indices as a robust tool for assessing the effluent stressed ecosystem in Nandgaon coastal waters, Maharashtra, India. Estuarine, Coastal and Shelf Science. 2014;**146**(5):60-67. DOI: 10.1016/j.ecss.2014.05.024

[105] Sivaraj S, Murugesan P, Silambarasan A, Preetha Mini Jose HM, Bharathidasan V. AMBI indices and multivariate approach to assess the ecological health of Vellar-Coleroon estuarine system undergoing various human activities. Marine Pollution Bulletin. 2015;**100**:334-343. DOI: 10.1016/j.marpolbul.2015.08.028

[106] Giangrande A, Licciano M, Pasqua M, Fanizzi F, Migoni D, Stabili L. Heavy metals in five Sabellidae species (Annelida, Polychaeta): Ecological implications. Environmental Science and Pollution Research. 2017;**24**:3759-3768. DOI: 10.1007/s11356-016-8089-8

[107] Otero X, Sánchez J, Macías F. Bioaccumulation of heavy metals

in thionic fluvisols by a marine polychaete: The role of metal sulfides. Journal of Environmental Quality. 2000;**29**:1133-1141. DOI: 10.2134/JEQ2000.00472425002900040014X

[108] Reish D, Martin J, Piltz F, Word J. The effect of heavy metals on laboratory populations of two polychaetes with comparisons to the water quality conditions and standards in Southern California marine waters. Water Research. 1976;**10**:299-302. DOI: 10.1016/0043-1354(76)90170-6

[109] Pires A, Velez C, Figueira E, Soares AM, Freitas R. Effects of sediment contamination on physiological and biochemical responses of the polychaete *Diopatra neapolitana*, an exploited natural resource. Marine Pollution Bulletin. 2017;**119**(1):119-131. DOI: 10.1016/j.marpolbul.2017.03.014

[110] Penha-Lopes G, Costa P, Gil J, Cannicci S, Macia A, Mwangi S, et al. Polychaeta community in peri-urban equatorial and subtropical mangroves of East Africa. Western Indian Ocean Journal of Marine Science. 2013;**12**:1-14. DOI: 10.4314/WIOJMS.V12I1

[111] Sivakumar P, Srinivasan G, Janardhanam M, Sivakumar R, Niranjani Marcus P, Balasubramaniam S, et al. A rapid bioassay test for assessing environmental contamination using the marine sedentary polychaete hydroides elegans. Water. 2022;**14**(11):1713. DOI: 10.3390/w14111713

[112] Keerthika K, Padmavathy P, Rani V, Jeyashakila R, Aanand S, Kutty R. Evidence of microplastics in the polychaete worm (capitellids—*Capitella capitata*)(Fabricicus, 1780) along Thoothukudi region. Environmental Monitoring and Assessment. 2024;**196**(6):1-13. DOI: 10.1007/s10661-024-12688-x

[113] Suriya J, Bharathiraja S, Sekar V, Rajasekaran R. Metallothionein induction and antioxidative responses in the estuarine polychaeta *Capitella capitata* (Capitellidae). Asian Pacific Journal of Tropical Biomedicine. 2012;**2**(2):S1052-S1059. DOI: 10.1016/S2221-1691(12)60360-8

[114] Lamont PA, Gage JD. Morphological responses of macrobenthic polychaetes to low oxygen on the Oman continental slope, NW Arabian Sea. Deep Sea Research Part II: Topical Studies in Oceanography. 2000;**47**:9-24. DOI: 10.1016/S0967-0645(99)00102-2

[115] Sarathy PP, Bharathidasan V, Gnanamoorthy P, Murugesan P. Influence of physico-chemical parameters and pCO2 concentration on mangroves-associated polychaetes at Pichavaram, southeast coast of India. SN Applied Sciences. 2019;**1**:1-13

[116] Martinez-Garcia E, Sanchez-Jerez P, Aguado-Giménez F, Ávila P, Guerrero A, Sánchez-Lizaso JL, et al. A meta-analysis approach to the effects of fish farming on soft bottom polychaeta assemblages in temperate regions. Marine Pollution Bulletin. 2013;**69**:165-171. DOI: 10.1016/j.marpolbul.2013.01.032

[117] Velvizhi S, Gopalakrishnan A, Murugesan P, Kannan D. A case study on polycheate fishery by the irular tribal fishing community on the Tamil Nadu Coast. Aquaculture Asia. 2013;**XVIII**(4):17-20

[118] Muthuvelu S, Murugesan P, Muniasamy M, Vijayalakshmi S, Balasubramanian T. Changes in benthic macrofaunal assemblages in relation to bottom trawling in Cuddalore and Parangipettai coastal waters, southeast coast of India. Ocean Science Journal. 2013;**48**(2):183-195. DOI: 10.1007/s12601-013-0015-0

[119] Danovaro R, Nepote E, Martire ML, Ciotti C, De Grandis G, Corinaldesi C, et al. Limited impact of beach nourishment on macrofaunal recruitment/settlement in a site of community interest in coastal area of the Adriatic Sea (Mediterranean Sea). Marine Pollution Bulletin. 2018;**128**:259-266. DOI: 10.1016/j.marpolbul.2018.01.033

[120] Omena EP, Lavrado HP, Paranhos R, Silva TA. Spatial distribution of intertidal sandy beach polychaeta along an estuarine and morphodynamic gradient in an eutrophic tropical bay. Marine Pollution Bulletin. 2012;**64**(9):1861-1873. DOI: 10.1016/j.marpolbul.2012.06.009

Chapter 4

Perspective Chapter: Mixed Shrimp Farming and Mangrove Landscapes – Between Myth and Reality

Gerardo Rodríguez-Quiroz, Walter Hubbard-Zamudio and Pedro Joaquín Gutiérrez-Yurrita

Abstract

The concept of landscape is polysemic and subjectively constructed. It comprises abstract and physical elements that shape landscape character and community's identity. Recognizing a natural link between nature and community does not create the landscape; it makes people feel dependent on their environment that provides them with services. The landscape creates an unequal relationship of "co-dependence" between people and territory. Humanity has devised many rural productive landscapes, whether emulating natural processes or removing environmental constraints that might impede production. Shrimp in mangrove ecosystems have been cultivated in Sinaloa for over five centuries. However, it was only in the last decades that they were transformed into semi-intensive and intensive shrimp farming. The current yield is significantly higher than that obtained from shrimp fishery. An integrated landscape perspective of shrimp farming in mangrove ecosystems facilitates understanding of the advantages that both systems currently derive from each other. Restoring a mangrove to fully natural conditions is impossible; further fragmenting the mangrove is unacceptable; but creating a coherent landscape that conserves the mangrove with human activities is desirable and feasible. The objective of this work was to determine how mangrove-shrimp farm landscape helps each other to exist in a world where both are vulnerable.

Keywords: mangrove habitats, mangrove biodiversity, geographical data, landscape theory, shrimp farming

1. Introduction

The environmental consequences of shrimp farming have been a subject of considerable debate. The primary issue that has been widely discussed is the destruction of mangrove forests to convert them into ponds. A review of literature from non-governmental organizations (NGOs) makes it evident that shrimp aquaculture is one

IntechOpen

of the primary causes of mangrove destruction on a global scale. These environmental groups posit that the rapid and unregulated expansion of shrimp farming in export-oriented developing countries has resulted in a state of environmental neglect that could potentially be catastrophic.

In the State of Sinaloa, there are a total of 907 crustacean production systems (CPS) served by the Sinaloa Aquaculture Health Committee (CESASIN in Spanish). This figure represents 7304 culture ponds of various types and sizes, installed in 67,345.63 ha of water surface. Of the total number of production systems, six are classified as extensive, 862 as semi-intensive, 25 as intensive, and nine as hyper-intensive [1]. The average number of production units in Sinaloa is 875 shrimp farms. The principal municipalities in which these farms operate are Ahorne, Guasave, Angostura, Navolato, Culiacán, Elota, Mazatlán, El Rosario, and Escuinapa [2].

The State of Sinaloa has undergone a significant transformation in its economic activities. Traditionally, the population was engaged in fishing and agricultural activities. However, in recent years, there has been a notable shift toward the development of a tourist zone. This has resulted in the alteration of the natural soil of the municipality, with urban land uses emerging that are more economically profitable for the owners [3].

The expansion of aquaculture areas in the state has resulted in the growth of mangrove forests in the coastal zone. The reforestation of coastal areas provides benefits to society by increasing the amount of open green space. However, this development effort is not optimal [4]. In a recent study, Hubbard-Zamudio, Rodríguez-Quiroz, and Gutiérrez-Yurrita [5] presented evidence that shrimp farming is not responsible for mangrove deforestation. Furthermore, the researchers found that the development and health of mangrove forests are not necessarily compromised by the presence of shrimp farms. Instead, the two can coexist, creating new bio-landscapes.

1.1 The practice of shrimp farming in Mexico

The contributions of the aquaculture sector have demonstrated that this activity is a significant promoter of global food security due to the diversity of food production, its substantial economic benefits, and the high generation of direct and indirect employment. However, there is still a notable deficit of knowledge from a social and conservationist perspective [6].

Shrimp aquaculture in Mexico represents a significant contribution to the national economy, exceeding fishery production by 75%. According to the latest data from the SADER 2022 report [7], there are 253,924 registered controlled production systems, with a live weight production of 198,067 T. Sinaloa, accounting for 46% of the total shrimp production, is the leading state within the Republic of Mexico. Sonora follows with 37% of the total production, and Nayarit is in third place with 6%. These states are all situated along the coast of the Mexican Pacific Ocean and have a latitude of above 20° N.

It had been anticipated that the growth rate would be more than 10%; however, due to a dearth of economic and technical resources in support of the sector—combined with the absence of suitable cultivation areas and an absence of a strategy for integrating shrimp farming with mangrove conservation—the rate has declined considerably, to 8.1% over the past decade. However, the latest statistical exercise of the SADER [7] indicated that the production of shrimp in aquaculture was 198,067 T, representing a decrease of 7.068% compared to 2020, which had the highest production figures. It is noteworthy that only the state of Sinaloa demonstrated an increase in shrimp aquaculture production during the same period, with a total production of 117,569 T [7].

At the present time, shrimp is the third most produced aquatic product in Mexico. However, in terms of economic revenue, it is in the first position, with figures exceeding $1,099,086,856 USD[1] [7]. The cultivation of the white shrimp (*Litopenaeus vannamei*, synonym *Penaeus vannamei*) is of particular significance in terms of its contribution to the family and regional economy [6]. The latest FAO statistical yearbook (2023) [8] highlights a concerning and urgent need for attention to the rise in malnutrition not only globally but also in countries such as Mexico, where high morbidity rates are attributed to inadequate nutrient intake and excess consumption of sugars and synthetic compounds that contribute to obesity across the lifespan. In this context, shrimp biotechnology emerges as a promising avenue for local economic development and enhanced community well-being [9].

It is paradoxical that despite Mexico having the environmental, economic, and social conditions to maintain or exceed international trends, official figures demonstrate limited growth in the sustainable use of our seas and coasts [10]. This is due to several underlying causes. As previously stated, it is imperative to implement a National Strategy that integrates shrimp farming with mangrove conservation, as both represent a significant economic resource on a global scale and a vulnerable ecosystem that requires protection [5, 11].

1.2 The conservation of mangrove forests

A prominent discourse posits that the aquaculture industry, and shrimp farming in particular, constitutes a significant driver of mangrove degradation. Mangroves are among the most productive ecosystems in the world, playing a vital role for the coasts from an ecological, economic, and social perspective [12]. The mangroves perform a range of environmental services, including the reduction of wind force and wave erosiveness, as well as the protection of inland facilities from the damaging impacts of storms, storm surges, and hurricanes. Furthermore, they play a vital role in maintaining seawater quality within estuaries, functioning as natural purifying filters [13, 14].

In recent years, the shrimp farming sector has made efforts to find a possible solution that would allow this highly lucrative industry to operate without damaging the mangrove ecosystem [5, 15]. Sinaloa's mangroves form a large wetland corridor and provide several ecosystem services; however, various stressors could increase their vulnerability and reduce the biological diversity of the ecosystem [16].

1.3 The dilemma

Mangroves occupy a small proportion of the country's land area, but as an ecosystem, they provide most ecosystem services. At the same time, it can be said that it is an indispensable semi-natural landscape for the sustainable and integral development of Mexico's coasts.

Among the coastal ecosystem services, mangroves are considered one of the most important ecosystem providers, as they provide habitat, refuge, and nursery for wildlife, since mangroves are habitat for protected species and for commercially important species; this last service is perhaps the most valued due to its positive relationship with fishing and other productive activities that can be carried out in a sustainable and harmonious manner with the landscape [17]. This does not mean that the problems are solved; on the contrary, aquaculture is well regulated in terms

[1] The official currency exchange rate on August 26, 2024, was $1USD = 19.40 MxP.

of environmental management on paper, in terms of legislation, but administratively there is a lack of specific measures and constant monitoring for each farm. Some measures to mitigate environmental impacts have been effective, but problems still occur. However, it must incorporate key management elements to avoid controversy in its implementation; citizen participation and alliances with external actors must be among the most important, as these elements would give public value to the objectives to be achieved with management [18].

A landscape approach is a possible alternative for optimizing the creation of environmental policies that become public in a conjunction of subjects, given that environmental policies are born in a territory that, when used by man, is transformed into a productive material, be it agricultural, forestry, or fisheries. Public policies in aquaculture, as in agriculture, help us to understand that landscapes should be understood as a paradox in the sense that it seems a contradiction the conservationist objective with respect to the productive objective, when in fact we have a dilemma of how to harmonize environmental protection with sustainable use in a development model that integrates social, economic, and environmental factors in the same space and time, providing tools for the allocation and management of land where different activities compete for its use [19].

Rentería, Sánchez, and Gutiérrez-Yurrita [20] propose a management based on the integrated landscape approach, adopting landscape theory in a pragmatic way under a conciliatory vision that considers the interested parties and/or actors and their needs through their active participation and processes of negotiation, planning, and management to increase capital [21]. In this instrumental approach to territorial management, the potential impact and likelihood of socio-environmental risks are strategically assessed using modern environmental law tools such as strategic landscape assessment (SLA) or strategic environmental assessment (SEA) [22]. A landscape-focused policy is cross-cutting, in contrast to the territorial approach, which is more sectoral, considering only one sector and largely ignoring possible impacts on other sectors such as agriculture, fisheries, urban or tourism, proposing one-dimensional solutions to multilateral problems [23].

Considering the aforementioned framework, our objective was to put forth the integrated landscape approach as a potential alternative for enhancing the formulation of productive public policies in aquaculture. This approach is based on three fundamental axes: (i) the landscape as an integrating and conciliatory approach; (ii) the incorporation of the landscape in public land-use planning policies; and (iii) an examination of the potential benefits and challenges associated with implementing a landscape approach in aquaculture public policies.

Nevertheless, the implementation of an integrated landscape approach necessitates the establishment of a collaborative framework between the relevant authority and the surrounding community. This framework should facilitate the development of aquaculture in the coastal mangrove forest, thereby enhancing the community's economic, social, and environmental well-being. Additionally, it should contribute to disaster mitigation strategies that promote sustainability and resilience [4].

Mitigation of environmental effects and land use planning are conducted in conjunction with a single activity, thereby rendering the concept of aquaculture sustainability contingent upon the environmental impact of associated activities and the adequacy of their development [24]. In alignment with the landscape approach, it is evident that both mangroves and aquaculture ponds provide cultural ecosystem services to local communities, including those primarily engaged in fishing, fish farming, and agriculture. The term "cultural ecosystem services" has been defined as a category of goods derived from natural systems which support a wide range of cultural and productive habits [25]. However, there is no literature that excludes these services in cases where natural systems receive some form of anthropogenic assistance or support. This is evidenced by a Collaboration Agreement on Productive Linkage-Technological Development signed with the Centro de Investigación en Alimentación y Desarrollo, A.C. (CIAD) and National Commission of Protected Natural Areas, although there is no scientific documentation to support this. In the 2019 financial year, the project entitled "Assisted recruitment for the repopulation of white shrimp (*Litopenaeus vannamei*) in the Huizache-Caimanero lagoon system' was carried out in collaboration with CIAD and National Fisheries Commission. In a coastal lagoon system that was once highly productive and is now in sharp decline, juvenile shrimp obtained in commercial laboratories of this species were planted.

2. Methodology

2.1 Normalized difference vegetation index (NDVI)

The NDVI helps differentiate vegetation from other types of land cover (artificial) and assess its overall condition. It also allows for the definition and visualization of vegetated areas on a map, as well as the detection of abnormal changes in the growth process [26]. NDVI is calculated using the following formula:

$$\text{NDVI} = (\text{NIR} - \text{red}) / (\text{NIR} + \text{red}) \tag{1}$$

where NIR is near-infrared light, and red is visible red light.

It works by mathematically comparing the amount of visible red light absorbed and near-infrared light reflected. This occurs because chlorophyll in a healthy plant absorbs most of the visible red light, while the cellular structure of the plant reflects most of the near-infrared light. This means that high photosynthetic activity, typically associated with dense vegetation, will result in less reflectance in the red band and more reflectance in the near-infrared band. By observing how these values compare, one can reliably detect and analyze vegetation cover separately from other types of natural land cover.

The Landsat 8 satellite images used to classify mangrove vegetation, as well as to digitize farms and key coastal features, formed the basis of the study. Changes in mangrove extent due to natural causes were considered negligible compared to the impact of shrimp pond installation. Therefore, the process of classifying land covers was initially conducted with images from 2013, prior to current aquaculture development, and compared with updated images from 2020. For this purpose, images with a resolution of 10 m were used.

2.2 Land surface temperature (LST)

Land surface temperature is a key environmental variable. LST represents a fundamental determinant of the Earth's thermal behavior, exerting control over the effective radiative temperature of the Earth's surface. In essence, LST can be defined as the heat that the Earth's surface would experience at a specific location. From the perspective of a satellite, the term "surface" refers to what is observed when the atmosphere is penetrated, and the ground is visible (**Figure 1**).

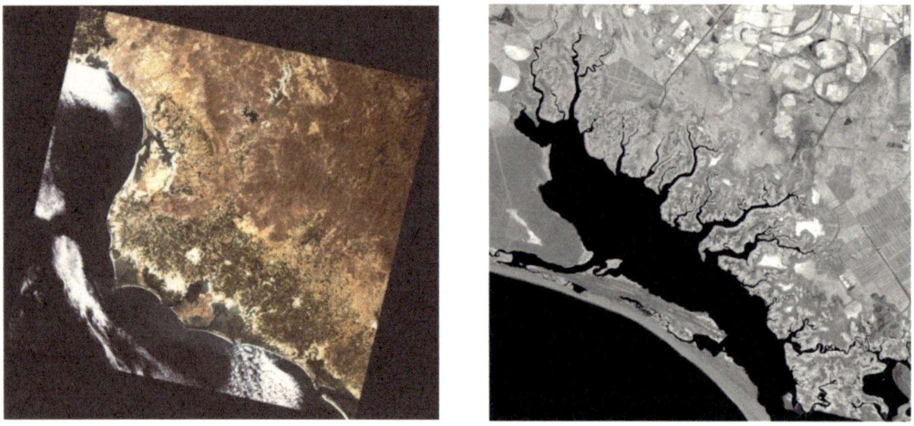

Figure 1.
Primary satellite imagery for LST estimation.

The surface in question could be snow and ice, the grass on a lawn, the roof of a building, or the leaves in a forest canopy. It should be noted that the LST is not synonymous with the air temperature that is typically included in daily weather reports (European Space Agency, ESA).

The alteration of land surface temperature is contingent upon a multitude of factors, including modifications in land utilization, land surface characteristics, seasonal fluctuations, climatic circumstances, and economic advancement [24].

For example, alterations in land cover associated with urbanization modify the microclimate, resulting in the formation of urban heat islands, whereby cities become warmer than the surrounding countryside. A comprehension of the factors that precipitate this phenomenon could assist urban areas in adapting to climate change and enhancing the quality of life for their inhabitants [24, 27].

Urban areas are distinguished by a reduction in vegetation cover and an expansion of impervious surfaces, including pavements, roads, and buildings, relative to rural areas. This urbanization contributes to higher temperatures in cities than in rural regions; indeed, several factors have been identified that can serve to mitigate the urban heat island effect. Among these factors, the influence of natural elements, including vegetation and water bodies, has been observed to be particularly significant [28, 29]. In this context, the conversion of marshland for aquaculture farms specializing in shrimp farming can be viewed as a parallel phenomenon.

Satellite measurements of sea surface temperature (SST) provide unique data sources for regional and global coverage with fairly good temporal and spatial resolution [30]. LST can alter or modify the temperature of the air in a lower layer or atmosphere. To calculate LST, the USGS formulas were used (algorithm for automated mapping of land surface temperature using Landsat 8 satellite data).

It is considered that using LST values from a single Landsat scene to represent an entire season or year shows some limitations in the representativeness of LST values due to daily temperature fluctuations. Thus, the analysis presented here is referential and aims to demonstrate that this type of tool can also help identify benefits related to ecosystem services derived from aquaculture activities.

1. TOA (Top of Atmospheric) Spectral Radiance Calculation

   TOA (Lλ) = ML * Qcal + AL
   Where:
   ML = Band-specific multiplicative rescaling factor (value available in the MTL metadata file, in the line RADIANCE_MULT_BAND_x, where x is the band number).
   RADIANCE_MULT_BAND_10 = 3.3420E-04.
   RADIANCE_ADD_BAND_10 = 0.10000

2. Conversion of TOA to Brightness Temperature

   Where:
   K2 and K1 = Band-specific thermal conversion constants from the metadata.
   K1_CONSTANT_BAND_10 = 774.8853.
   K2_CONSTANT_BAND_10 = 1321.0789

3. NDVI Calculation

   NDVI = (Band 5 - Band 4) / (Band 5 + Band 4)

4. Considering that NDVI calculation is important because, subsequently, the proportion of vegetation (Pv) must be calculated, which is highly related to NDVI, and the emissivity (ε), which is related to Pv.

5. NDVI = Float(Band 5 - Band 4) / Float(Band 5 + Band 4)

6. Vegetation Proportion Calculation (Pv)

   Pv = Square ((NDVI - NDVImin) / (NDVImax - NDVImin))

7. Emissivity Calculation (ε)

   ε = 0.004 * Pv + 0.986.
   The value of 0.986 corresponds to a correction value in the equation.

8. Land Surface Temperature Calculation

   LST = (BT / (1 + (0.00115 * BT / 1.4388) * Ln(ε))).

Figure 2 shows the final image that is obtained at each of the steps in the process of obtaining the LST.

2.3 Shrimp farming infrastructure

The study area is situated in the northern region of Sinaloa, Mexico, with coordinates of 25°36'N, 109°06'W. The Ramsar Site number 107 on the national list and 2025 on the Ramsar list, known as "Lagunas de Santa María (4,000 ha)-Topolobampo (6,000 ha)-Ohuira (12,500 ha)" is situated within the study area. The original polygon representing the study area was obtained as a KML file from the webpage [31].

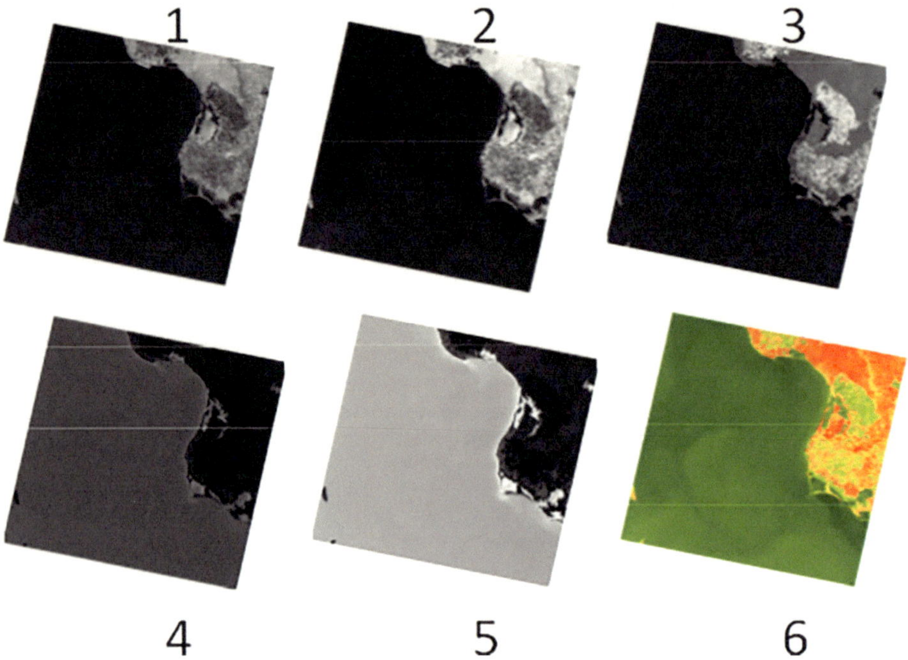

Figure 2.
Steps to follow to obtain LST.

A land cover classification map of the study location was generated using Landsat satellite imagery, with consideration given to the distribution of hectares of mangrove and non-mangrove vegetation (**Figure 3** upper).

The study site is a mangrove forest, primarily characterized by the presence of three main mangrove species: black mangrove (*Avicennia germinans*), white mangrove (*Laguncularia racemosa*), and red mangrove (*Rhizophora mangle*). These species are considered to have significant ecological importance and are thus subject to "Special Protection" status (NOM-059-SEMARNAT, 2010; Official Mexican Standard 059 of the Secretariat of Environment and Natural Resources).

Direct land use in shrimp farming encompasses a range of activities, including the construction of production ponds (nursery and grow-out), the creation of reservoirs to retain water for a few hours or days before transferring it to production ponds, the establishment of channels for intake and discharge, the development of effluent sedimentation basins, the construction of embankments and space for buildings, preparation, and open-air storage. The referenced polygonal areas of shrimp farms operating in the study area were obtained from satellite images and grouped as blocks.

2.4 The existing infrastructure within the area of study

The existing infrastructure within the area of study is examined to gain insight into the current situation. The current areas were divided into three bloques separated by the swamp, where there is no mangrove vegetation in the lateral part separating these bloques. Bloque 1, with farms located in the distal part of the north, Bloque 2, which corresponds to the infrastructure grouped in the central part of the study area, and Bloque 3, with farms located in the extreme south (**Figure 3** lower).

The selected area encompassed a total surface area of 2688.6 hectares dedicated to shrimp ponds. The total surface area of the water inlet channels was 161.084 hectares, while the harvest channels, which were identified as potential sites for deforestation, encompassed 215.022 hectares. Two annual crop cycles were proposed to produce approximately 3200 tons of shrimp of varying sizes and weights, representing the total output of the identified farms within the study area.

The initial contact between the structure and sea water, which may eventually result in the deforestation of mangroves, is through the water intake channels, excavations, and extensions used for pumping estuarine water to the farm. This water is initially used for filling the pond and then for replenishments that maintain the water quality within it. The calculated surface area for this infrastructure is 161.1 hectares. The harvesting channels, which were designed to have the same characteristics as the

Figure 3.
The study area is illustrated with three distinct working blocks (estuarine complex), each of which contains a shrimp farm.

calling channel, were generally conceived to connect with the existing natural channels that flow into the coastal lagoon system (**Figure 1** lower). The length and characteristics of these channels were largely dependent on the specific geography of the site in question. The calculated surface area for this infrastructure was 215 hectares.

3. Commented results

The referenced polygonal areas of shrimp farms operating in the study area were obtained from satellite imagery. In 2013, considered as the first reference year, it presented a total area of 1792 ha, divided into two bloques (1 and 3, ordered from north to south), with no aquaculture infrastructure between them. The comparative reference year was 2020, with a total area of 3065 ha, divided into three bloques. This represents a growth of 71% in 7 years (**Table 1**).

The current climate trends, as well as their impact on vegetation cover, have motivated a variety of research projects worldwide. Infrared thermal sensors have been successfully applied in environmental and climate studies, mainly to analyze surface temperature patterns and relate them to landscape characteristics [24, 28]. In this regard, alterations in land use/cover and land surface temperatures play a pivotal role in the genesis and intensity of urban heat island (UHI) effects [29, 32]. The final statistics of the LST images, captured at two different times (2013 and 2020), are presented in **Table 2**. The resulting images are displayed in **Figure 4**.

In 2020, the average image value was 28.84. This value was found to be slightly lower (29.01) for blocks with farms than for the mangrove zone (29.23), with a differential of 0.22. It can thus be concluded that areas with farms exhibited a slight decrease in temperature of 0.22 compared to areas without farms. The following satellite images of LST have been compiled to facilitate a comparative analysis between the shrimp pond bloques and the surrounding natural mangrove and estuarine areas (**Figure 5**).

The present study sought to identify the temperature existing in the areas where the aquaculture infrastructure was constructed, as well as the temperature in the immediately adjacent area, from the perspective of the formation of "heat islands". The results indicate that there were no significant increases in temperature in the area where the aquaculture infrastructure was constructed. Furthermore, although the observed values are not statistically significant enough to determine a difference between the two areas, they do present lower values in both years under evaluation. Our results support the theory that water bodies are a mitigant to temperature increases in agricultural and urban areas [24, 27, 32, 33].

Bloque/Reference year	2013	2020
Bloque 1	270 ha	361 ha
Bloque 2	—	419 ha
Bloque 3	1522 ha	2285 ha
Total	1792 ha	3065 ha

Table 1.
Growth in area (hectares) of shrimp farms in the study area in 7 years.

Year	Average	Highest	Lowest	Sd
2013	22.64	41.18	14.70	5.20
2020	28.85	31.52	21.67	1.75

Table 2.
LST image statistics from 2013 to 2020.

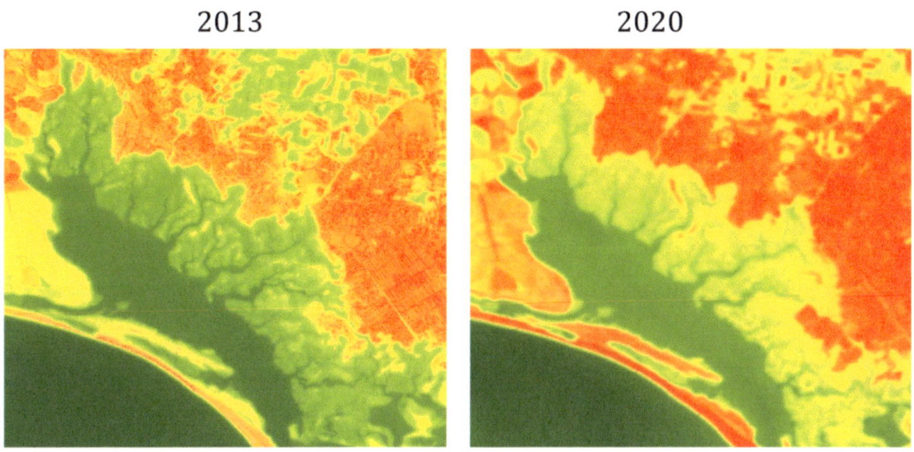

Figure 4.
Final LST satellite images for comparative analysis between 2013 and 2020.

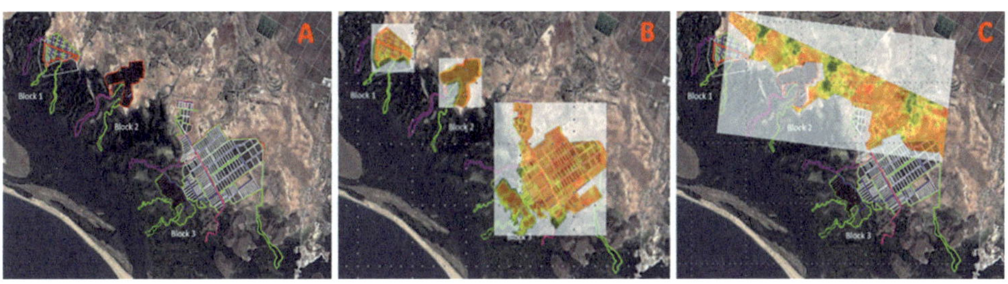

Figure 5.
Satellite images of LST compared between Bloques with shrimp ponds and natural mangrove and estuarine areas. A. Satellite image showing the current shrimp farming infrastructure; B. Image with heat spots in shrimping Bloques zone; C. Image with heat spots in mangrove and estuary lands.

In their 2022 study, Cuthbert et al. [34] present results that lead to the important practical implication that irrigation in more arid locations can hypothetically improve the relative potential cooling performance while maintaining substantial retention performance. However, in many arid areas, the availability of a sustainable source of irrigation water may be a fundamental constraint to the applicability of such urban greening interventions as a cooling strategy. In our case, this irrigation substitution would be related to the daily filling and discharge of shrimp ponds from the coastal lagoon system that supplies it.

The present work is not a direct response to a particular question or problem; rather, it is an opening for new questions based on scientific scenarios that promote the use of scientific alternatives. In this case, the alternative in question is the phenomenon of "heat islands," which have been proposed as a negative environmental indicator. In this study, we propose that they can also be used as a reference point for improving the environmental sustainability of certain productive activities by contributing to the provision of ecosystem services in sensitive or vulnerable landscapes.

4. Discussion

The concept of landscape is polysemic, constructed subjectively with abstract elements identified with the personal character of each person and a material perspective that concretizes it in a specific place and time. Although it is acknowledged that there is a natural human connection, an epigenetic one might suggest, to a territory and the environmental setting it comprises, this primitive identification does not create the landscape. Rather, it simply recreates a sense of human dependence on an environment that provides them with services, both tangible and intangible. The formation of the landscape is now recognized as a process whereby humanity establishes a symbiotic relationship with the territory in which it is situated, where it has learned to function and develop its identity [35].

The current state of agricultural and aquacultural production is characterized by a high degree of diversity, with practices ranging from the emulation of natural ecosystems to the manipulation of environmental factors that could otherwise impede productivity. The semi-intensive cultivation of shrimp in mangrove ecosystems is a notable example of this evolution, with developments now focused on maintaining ecological coherence in the face of climate change and the heightened vulnerability of mangrove ecosystems to external disturbances.

In the context of our study, humanity has developed productive landscapes within a distinctive vegetation zone that is susceptible to significant climatic and structural alterations in the ecosystem. Mangrove landscapes with shrimp farming represent an integrated approach to land use. Both systems are mutually reinforcing, providing the stability that mangroves have historically lacked due to other ecological modifications that humans have introduced to the coast, the rivers that flow into the estuary, and the urban settlements within the hydrographic basin.

In the mangroves of the Mexican state of Sinaloa, shrimp farming has been a rapidly expanding industry since 2013. The volumes reached in a decade exceeded those traditionally obtained from wild fishing. As previously indicated, shrimp farming is the predominant source of shrimp production, accounting for over 70% of the total, while wild fishing contributes only 20% on average over the past decade. Sinaloa remains the largest producer, holding a share of 40.3–46% of the national total [7].

Aquaculture production in Sinaloa encompasses approximately 40,000 hectares of aquatic surface. From an economic perspective, this production area is equivalent to a harvest of approximately 99,000 tons of white shrimp (*L. vannamei*) with an estimated value of 500 million dollars [7]. However, the development of shrimp infrastructure continues to be associated with the destructive conversion of mangroves, including physical destruction (forestry, draining of wetlands, and coastal transformation).

The initial clearance of mangroves to create water channels, ponds, laboratory facilities, and production areas was a one-time occurrence. However, this initial phase of shrimp farming has since evolved into a more permanent and extensive operation. It is evident that shrimp farming cannot be 100% sustainable or ecological, given that it is an anthropogenic activity that has become increasingly intensified, technological, and situated within increasingly populated landscapes. However, it can be an activity that harmonizes tradition, culture of sustainable use, and conservation of nature in the small coastal communities of Sinaloa (an approach that could be extended to all of Mexico with the requisite prior studies).

Therefore, for integrated management, we consider it useful to relate shrimp farming and mangroves under landscape management scenarios, as described by the European Landscape Convention as "any part of the territory perceived by the population, whose character is the result of the action and interaction of natural and/or human factors" [36], currently modernized by focusing the landscape on increasing the relationships between individuals and societies.

In the present analysis, the term "mangrove" is used to refer to a group of wetlands dominated by trees that colonize the intertidal zone. Mangroves are plant communities that occur throughout the world along tropical and subtropical coasts, situated at the convergence of land and sea. This unique location creates a new and complex scenario for humans. Its emergent properties render it more resilient due to the creation of internal subsystems that make it less vulnerable. However, this resilience is accompanied by increased sensitivity to long-term human disturbances, as degradation will necessitate significant restoration or rehabilitation efforts [37].

At different times (1999, 2011 for example), it has been estimated how much mangrove has been lost in the Mexican coasts, especially in Sinaloa, presumably due to aquaculture activities [38, 39]; nevertheless, no conclusive evidence exists that the reduction of mangrove forests is solely attributable to the expansion of shrimp farming infrastructure [5].

The loss or gain of mangrove area is influenced by several factors, some of which are directly related to, and some of which are not, to shrimp farming [40]. The present study demonstrated the initial extent of mangrove cover prior to the construction of aquaculture infrastructure and the subsequent reduction in this area resulting from the development of the infrastructure.

The hypothesis that the development of coastal shrimp farms would result in the deforestation of mangrove areas was tested using modern geographic information techniques, landscape ecology, and remote sensing analysis. The aim was to investigate whether the modification of mangrove areas was due to specific actions taken during the construction of shrimp farms in an aquaculture development hub in northwestern Mexico.

It is imperative to assess the potential for expansion in shrimp farming. Can additional ponds, channels, and other infrastructure be constructed? It is probable that the landscape system is already optimized and can be managed sustainably; however, this is not a limitless scenario, as every landscape has physical limits. The landscape is in a constant state of dynamic change, and the natural evolution of the landscape cannot be halted.

5. Conclusions

Although the use of a single Landsat scene may not capture daily temperature fluctuations, the results offer valuable reference data to demonstrate the potential of LST as a tool for ecosystem monitoring. The findings suggest that integrating LST calculations with other environmental metrics can enhance our understanding of how aquaculture and other land use changes affect local climates and ecosystem services.

LST is a powerful indicator for assessing land use impacts, particularly in sensitive coastal regions where aquaculture development is prevalent. The ability to calculate LST from satellite data provides researchers and policymakers with critical information to guide sustainable land management and monitor environmental changes over time.

The loss or gain of mangrove surface area is influenced by several factors, some of which are directly related, and others not, to shrimp farming in the early days of its great development pole, many years ago.

Given the impracticality of reverting the mangrove ecosystem to a state approximating its original, it is crucial to pursue an integrated landscape perspective that incorporates both the human environment of shrimp farming and the natural ecosystem of mangroves. This approach allows for a comprehensive understanding of the mutual benefits that these two systems currently derive from one another.

The interdependence between the two systems commenced decades ago and has recently been reinforced. Consequently, we now possess a more comprehensive grasp of the interrelationship between them.

Integrated coastal landscape is a holistic approach combining the productive use of coastal ecosystems, such as shrimp farming or sustainable resource utilization, while conservating critical habitats like mangroves and marshes. Thus, the term would emphasize balancing economic activities and environmental stewardship to ensure long-term sustainability and resilience of these interconnected ecosystems.

Author details

Gerardo Rodríguez-Quiroz, Walter Hubbard-Zamudio
and Pedro Joaquín Gutiérrez-Yurrita*
National Polytechnic Institute, México

*Address all correspondence to: pgutierrezy@ipn.mx

References

[1] CESASIN. Available from: Sanidad acuícola en el Estado de Sinaloa. 2024. Available from: https://osiap.org.mx/senasica/sector-estado/sinaloa/Acuicola

[2] Gobierno de Sinaloa. Programa sectorial 2017-2021. Secretaría de Pesca y Acuacultura. El estado de Sinaloa: Órgano oficial del Gobierno del estado; 2018. Available from: https://media.transparencia.sinaloa.gob.mx/uploads/files/1/Programa%20Sectorial%202017-2021%20Pesca.pdf

[3] López, Araiza Hernández DP, Ramírez Sarabia JR. Pérdida del Mangle por el Mercado Inmobiliario turístico, el caso de Mazatlán, Sinaloa, México. In: Mora Cantellano PA, Serrano Oswald SE, Mota Flores VE, editors. Reconfigurando territorios a partir de la cultura, el empoderamiento de las mujeres y nuevos turismos. México: UNAM-AMECIDER; 2022. pp. 533-544 https://ru.iiec.unam.mx/5907/1/1.%20182-LopezAraiza-Ruiz.pdf

[4] Fuady M. Ecotourism concept to support ecological function of mangrove forest in coastal area of Banda Aceh city. In: AIP Conference Proceedings. Vol. 3082. 2024. p. 030034. DOI: 10.1063/5.0201080

[5] Hubbard-Zamudio W, Rodriguez-Quiroz G, Gutiérrez-Yurrita PJ. Shrimp farming and mangroves: Entities in conflict? Latin American Journal of Aquatic Research. 2023;**51**(5):642-657. DOI: 10.3856/vol51-issue5-fulltext-3057

[6] González GK, Peinado VM, Peinado HJ. Aporte al desarrollo económico y bienestar social de la actividad acuícola camaronera en el municipio Guasave, Sinaloa, México. In: Isaac Egurrola JE, Morales ER, Treviño Aldape A, editors. La economía sectorial reconfigurando el territorio y nuevos escenarios en la dinámica urbano rural. México: UNAM-AMECIDER; 2022. pp. 131-146. Available from: http://ru.iiec.unam.mx/5828/

[7] SADER (Secretaría de Agricultura y Desarrollo Rural). Anuario estadístico. 2022. Available from: https://nube.conapesca.gob.mx/sites/cona/dgppe/2022/ANUARIO_ESTADISTICO_DE_ACUACULTURA_Y_PESCA_2022.pdf

[8] FAO (Food and Agriculture Organization). Anuario estadístico. 2023. Available from: https://openknowledge.fao.org/server/api/core/bitstreams/28cfd24e-81a9-4ebc-b2b5-4095fe5b1dab/content/cc8166en.html

[9] Tacon AG, Lemos D, Metian M. Fish for health: Improved nutritional quality of cultured fish for human consumption. Reviews in Fisheries Science and Aquaculture. 2020;**28**:449-458

[10] Zambrano Salgado L, Gutiérrez-Yurrita PJ. Los mares mexicanos como socioterritorios sustentables y estratégicos para México. Revista Mexicana de Investigación-Interdisciplinaria. 2019;**2**(1):1-33 https://redasmiimx.org/numeros.html

[11] Cuéllar-Lugo MB, Asiain-Hoyos A, Juárez-Sánchez JP, Reta-Mendiola JL, Gallardo-López F. Evolución normativa e institucional de la acuacultura en México. Agricultura, Sociedad y Desarrollo. 2018;**15**(4):541-564

[12] Cerri F, Giustra M, Anadol Y, et al. Natural products from mangroves: An overview of the anticancer potential of Avicennia marina. Pharmaceutics. 2022;**14**(12):2793

[13] Lee SY, Primavera JH, Dahdouh-Guebas F, et al. Ecological role and services of tropical mangrove ecosystems: A reassessment. Global Ecology and Biogeography. 2014;**23**(7):726-743

[14] Cerri F, Louis YD, Fallati L, et al. Mangroves of the Maldives: A review of their distribution, diversity, ecological importance and biodiversity of associated flora and fauna. Aquatic Sciences. 2024;**86**(**2**):44-67

[15] Álvarez M, Ochoa E, Villalba A, Siu E. Perfil De Caracterización De Las Prácticas De Camaronicultura En Sinaloa. In: Prácticas de Desarrollo Sostenible en Ambientes Costeros de Prioridad de los Ecosistemas del Golfo de California Marinas Recreativas y Maricultura del Coastal Resources Center. USA: University of Rhode Island; 2001

[16] Millán-Aguilar O, Nettel-Hernanz A, Hurtado-Oliva MA, Dodd R, Flores-Cárdenas F, et al. Landscape metrics and conservation status of five mangrove wetlands in the eastern gulf of California margin. Journal of Coastal Research. 2020;**36**:94-102. DOI: 10.2112/JCOASTRES-D-18-00060.1

[17] Fernández-Díaz VZ, Canul Turriza RA, Kuc Castilla A, Hinojosa-Huerta O. Loss of coastal ecosystem services in Mexico: An approach to economic valuation in the face of sea level rise. Frontiers in Marine Science. 2022;**9**:898904. DOI: 10.3389/fmars.2022.898904

[18] Ramírez Valdez CJ. Evaluación de la gestión ambiental sobre la actividad acuícola en el municipio de Guasave, Sinaloa. México: Colegio de la Frontera Norte; 2010

[19] Agnoletti M. Rural landscape, nature conservation and culture: Some notes on research trends and management approaches from a (southern) European perspective. Landscape and Urban Planning. 2014;**126**:66-73. DOI: 10.1016/j.landurbplan.2014.02.012

[20] Rentería, Tamayo OG, Sánchez E, Gutiérrez-Yurrita PJ. El paisaje como concepto conciliador para la formulación de leyes y políticas públicas agrarias en México. Revisión del tema. Revista Aranzadi de Derecho Ambiental. 2024:59. In press

[21] Masnavi MR, Motedayen H, Saboonchi P, Hemmati M. Analyses of landscape concept and landscape approach from theoretical to operational levels: A review of literature. MANZAR, The Scientific Journal of landscape. 2021;**13**(**57**):22-37. DOI: 10.22034/MANZAR.2021.283818.2128

[22] Gutiérrez-Yurrita PJ. Los corredores ecológicos como herramienta para paliar los efectos negativos del cambio climático sobre la biodiversidad. Derecho Ambiental y Ecología. 2007;**4**(21):55-63 https://www.researchgate.net/publication/255950171_LOS_CORREDORES_ECOLOGICOS_COMO_HERRAMIENTA_PARA_PALIAR_LOS_EFECTOS_NEGATIVOS_DEL_CAMBIO_CLIMATICO_SOBRE_LA_BIODIVERSIDAD

[23] Langston JD, Riggs RA, Boedhihartono AK, Kastanya A, Sayer J. An island in transition: Governing conservation and development in Seram, Indonesia, Singapore. Journal of Tropical Geography. 2020;**41**(**3**):413-431. DOI: 10.1111/sjtg.12336

[24] Jiang J, Tian GJ. Analysis of the impact of land use/land cover change on land surface temperature with remote sensing. Procedia Environmental Sciences. 2010;**2**:571-575. DOI: 10.1016/j.proenv.2010.10.062

[25] Kovács B, Márquez MA, Rodríguez P, Gutiérrez-Yurrita PJ, Perez-Verdin G. Analysis of cultural ecosystem services of rock climbing settings in Mexico City: The case of Los Dinamos Recreational Park. Ecosystems and People. 2021;**17**(1):370-382. DOI: 10.1080/26395916.2021.1946594

[26] Manrique EG. Índice de vegetación. Aplicación del NDVI. In: Teledetección. Avances y Aplicaciones. VIII Congreso Nacional de Teledetección. Asociación Española de Teledetección. España: Albacete; 1999. pp. 217-219

[27] Ouyang L, Yang Y, Wu Z, Jiang Q, Qiao R. Towards inclusive urbanism: An examination of urban environment strategies for enhancing social equity in Chengdu's housing zones. Sustainable Cities and Society. 2024;**107**:105414. DOI: 10.1016/j.scs.2024.105414

[28] Alexander C. Normalised difference spectral indices and urban land cover as indicators of land surface temperature (LST). International Journal of Applied Earth Observation and Geoinformation. 2020;**86**:102013. DOI: 10.1016/j.jag.2019.102013

[29] Liu Y, An Z, Ming Y. Simulating influences of land use/land cover composition and configuration on urban heat island using machine learning. Sustainable Cities and Society. 2024;**108**:105482. DOI: 10.1016/j.scs.2024.105482

[30] Yu L, Wu Z, Zhang R, Yang X. Partial least regression approach to forecast the east Asian winter monsoon using Eurasian snow cover and sea surface temperature. Climate Dynamics. 2018;**51**:4573-4584

[31] CONANP. Available from: http://www.conanp.gob.mx/conanp/dominios/ramsar/lsr.php

[32] Kafy AA, Saha M, Rahaman ZA, Rahman MT, et al. Predicting the impacts of land use/land cover changes on seasonal urban thermal characteristics using machine learning algorithms. Building and Environment. 2022;**217**:109066. DOI: 10.1016/j.buildenv.2022.109066

[33] Zuluaga Gomez OA, Patino Quinchia JE, Valencia Hernandez GM. Models implemented in the land surface temperature and vegetation indexes time series analysis: A taxonomic proposal in the context of the global climate change. Revista De Geografia Norte Grande. 2021;**78**:323-344. DOI: 10.4067/S0718-34022021000100323

[34] Cuthbert RN, Diagne C, Hudgins EJ, Turbelin A, Ahmed DA, et al. Biological invasion costs reveal insufficient proactive management worldwide. Science of the Total Environment. 2022;**819**:153404. DOI: 10.1016/j.scitotenv.2022.153404

[35] Garibay-Castro LR, Gutiérrez-Yurrita PJ, Hernández-Ruíz J, Trejo-Espino JL. Potential distribution and medicinal uses of the Mexican plant *Cuphea aequipetala* Cav. (Lythraceae). Diversity. 2022;**14**(**5**):403. DOI: 10.3390/d14050403

[36] Déjeant-Pons M. The European landscape convention. Landscape Research. 2006;**31**(**4**):363-384. DOI: 10.1080/01426390601004343

[37] Twilley RR, Day JW. Mangrove Wetlands. Estuarine Ecology. 2a ed. Nueva Jersey: Wiley-Blackwell, John Wiley and Sons, Inc. Publications; 2013. pp. 165-202

[38] Ruiz-Luna A, Berlanga-Robles CA. Modifications in coverage patterns and land use around the Huizache-Caimanero lagoon system, Sinaloa,

Mexico: A multi-temporal analysis using LANDSAT images. Estuarine, Coastal and Shelf Science. 1999;**49**(1):37-44. DOI: 10.1006/ecss.1999.0489

[39] Berlanga-Robles CA, Ruiz-Luna A, Bocco G, Vekerdy Z. Spatial analysis of the impact of shrimp culture on the coastal wetlands on the northern coast of Sinaloa, Mexico. Ocean and Coastal Management. 2011;**54**:535-543. DOI: 10.1016/j.ocecoaman.2011.04.004

[40] Carrasquilla-Henao M, Ocampo HAG, González AL, Quiroz GR. Mangrove forest and artisanal fishery in the southern part of the Gulf of California, Mexico. Ocean and Coastal Management. 2013;**83**:75-80. DOI: 10.1016/j.ocecoaman.2013.02.019

Chapter 5

Ecological Factors Shaping Marine Actinobacterial Niches in Diverse Coastal Habitats of the Andamans Sea, India

Rajagopal Gobalakrishnan and Kannan Sivakumar

Abstract

Marine microbial communities are structured by the spatial temporal, variability of physico-chemical, and biotic parameters. These parameters form a useful tool for the ecological assessment and monitoring of coastal ecosystems. The present investigation aims to study the influence of the physico-chemical parameters on the distributional variations of culturable actinobacterial population density, at six different locations on the Havelock Island coast of the Andamans, India. The study covered the important parameters *viz.* temperature in air (29–23°C), surface water (26–24°C), sediments (27–23°C), salinity in water (18–35 psu), sediments (17–34 psu), pH in water (6–8) and sediments (7.6–8.7), dissolved oxygen (DO) (4.60–2.10 ml l^{-1}), electrical conductivity (EC) (2–5.5 dS m^{-1}), nitrogen (N) (0.55–8 mg g^{-1}), potassium (K) (2.63–10.88 mg g¯1), phosphorus (P) (4.50–3.13 mg g^{-1}), total organic carbon (TOC) (60.88–11.88 mg g^{-1}), sand (65–95%), silt (3.5–9.4%), and clay (1.2–30%). Culturable actinobacterial population density in the sediments ranged from 8×10^3 to 21×10^3 CFU g^{-1}, registering the higher value at station 4 (mangrove) and the lower value at station 3 (beach). Cluster and principal component analysis (PCA) results have revealed that the mangrove environment favors a higher density of actinobacteria than the coral, seagrass, and beach habitats of the Island.

Keywords: actinobacteria, Havelock Island, mangroves, microbial ecology, physico-chemical characters

1. Introduction

An ecosystem comprises the collective presence of organisms inhabiting a particular region, alongside the non-living elements of the environment, including air, soil, and water, that interact with these organisms. Especially, marine ecosystems are dynamic with valuable natural resources and are the largest among the Earth's aquatic systems. They have important environmental values and their destruction will affect the whole marine environment [1]. Therefore, the study regarding the physico-chemical properties of the coastal environment is very essential.

Mostly the marine ecosystem is fueled by the regeneration of nutrients, mediated by the marine microbes and these are key to the decomposition of organic matter and the reformation of nutrients, thereby determining the productivity of an ecosystem [2]. Microorganisms inhabit every corner of the ocean, with diverse habitats ranging from open waters, sediments, and the bodies of marine organisms to estuaries and hydrothermal vents [2], and they play a crucial role in marine processes, actively participating in the transformation and mineralization of organic materials within both sediments and overlying waters [3].

The structure of microbial communities is shaped by the temporal and spatial fluctuations of physico-chemical and biotic parameters. The distribution, diversity, and activities of microorganisms are influenced by a multitude of hydro-biological factors and nutrient levels present in the aquatic environment. Extensive research has been conducted in the marine environment to thoroughly investigate these relationships [4, 5]. The distribution of microbial diversity depends on changes in water and other physico-chemical parameters [6]. They play important roles in maintaining the ecosystems and replaying lay to environmental changes and anthropic activities [7]. In comparison to other benthic organisms, microbial communities exhibit remarkably rapid responses to environmental and pollution changes. Moreover, the fate of the coastal marine sediment deposits is entirely dependent on the activities of these microorganisms [8, 9].

For almost two decades, actinobacteria have been considered an important constituent of the microbial communities in marine ecosystems [10], particularly in marine sediments, where they can reach up to 13% of the bacterial abundance [11–14]. Actinobacteria are very widely distributed in the world's oceans and truly indigenous marine actinobacteria have now been described [15]. They play pivotal ecological roles in the carbon cycle with their capability to break down organic materials and recycle them as done by their terrestrial counterparts [16]. Exploration of marine actinobacteria will not only address fundamental questions in microbial ecology but also yield valuable insights for harnessing their biotechnological potential. However, there are still numerous unanswered queries concerning the ecological roles of actinobacteria in marine environments, as well as their distribution patterns and evolutionary origins. There is a growing recognition that marine actinobacteria represent a significant reservoir for the exploration and discovery of novel secondary metabolites with great potential for biotechnological applications [17].

The combined analysis of water and sediment physico-chemical characteristics serves as a powerful approach for ecologically assessing and monitoring diverse coastal ecosystems. Key factors, such as temperature, salinity, dissolved oxygen, and sediment properties, collectively exert significant influence on the abundance and distribution of benthic organisms. It is important to note that no single factor alone holds a dominant impact, as these variables operate in conjunction to shape the ecological dynamics [18, 19]. To ensure the high quality of coastal water resources, it is essential to assess the pollution loads by monitoring a range of physico-chemical parameters [20]. In light of this, the objective of the present study was to investigate how variations in the distribution of culturable actinobacterial population density in the coastal region of Havelock Island, Andamans, are influenced by these physico-chemical parameters.

2. Materials and methods

2.1 Study area

The Andaman group of islands is situated in the Bay of Bengal, spanning between latitude 11°97′N and longitude 93°00′E. For this study, Havelock Island, encompassing mangrove, coral, seagrass, and beach habitats, was chosen. The selection of sampling stations around this island was based on the various habitats present (**Table 1** and **Figure 1**).

S. No	Stations	Location	Habitats	Coordinates
1	Station 1	Boat jetty	Mangrove	12°02′27.9″N 92°58′41.5″E
2	Station 2	EL Dorado	Coral	12°01′31.6″N 93°00′13.8″E
3	Station 3	Dive India	Beach	12°01′38.5″N 93°00′15.8″E
4	Station 4	Kalapathar	Mangrove	11°57′37.9″N 93°00′46.6″E
5	Station 5	Silver sand	Seagrass	12°00′36.1″N 93°00′31.0″E
6	Station 6	Radha Nagar	Beach	11°59′04.2″N 92°57′04.0″E

Table 1.
The sampling stations around the Havelock Island.

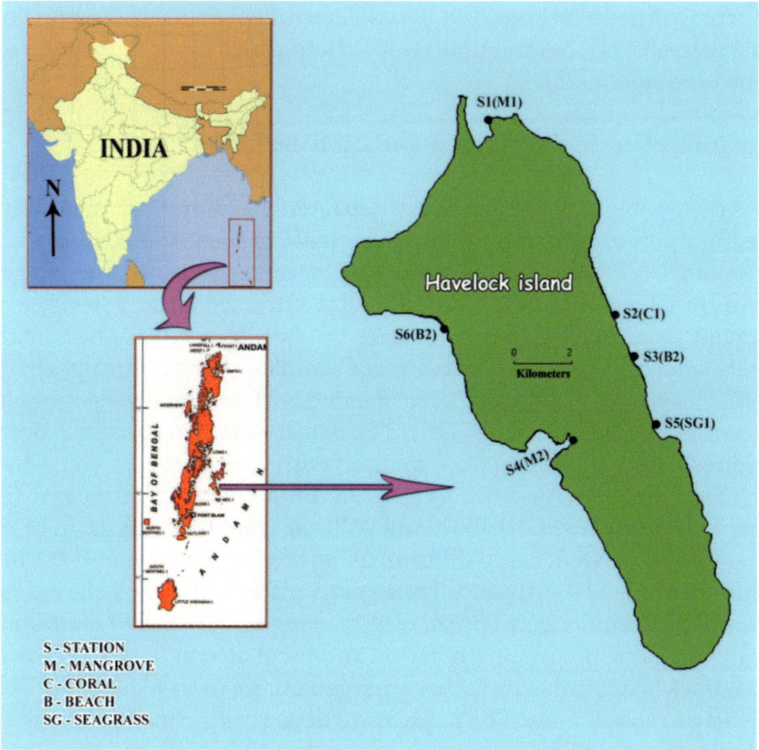

Figure 1.
Map showing the sampling stations in the Havelock Island.

2.2 Collection of samples

Field collections were conducted at six stations on Havelock Island to document diverse ecological characteristics (physico-chemical parameters), and collect sediment samples for actinobacteriological analysis. Using a sterile spatula, sediment samples were obtained at a depth of 25 cm from chosen locations on Havelock Island. These samples were carefully placed in sterile polythene covers and promptly transported to the field laboratory. Upon arrival, essential dilutions were prepared to facilitate subsequent microbiological analysis.

2.3 Physico-chemical parameters

Physico-chemical parameters *viz*. air temperature, surface water temperature, and sediment temperature were recorded using a mercury thermometer having ±0.02°C accuracy. The salinity of the water and sediments was recorded using a refractometer (Erma Company, Japan); Hydrogen-ion concentration (pH) of water was measured by a calibrated pH pen (pH Scan 1 Tester-Eutech Instruments, Singapore) and sediment pH was measured by a soil pH tester Model DM-13, Takemura Electric Works Ltd., Tokyo, Japan. Electrical conductivity (EC) was measured by Pye Unicam model 292 meters.

Water dissolved oxygen (DO) was estimated using the modified Winkler's method, described by Strickland and Parsons [21]. Sediment samples were used for the estimation of nitrogen, phosphorus and potassium, using Kjeldahl method [22] and colorimetric method [23], in the Sugarcane Breeding Research Institute, Cuddalore. Total organic carbon (TOC) was determined using potassium chromate as an oxidizing reagent [24]. Sediment texture was assessed using the pipette method, described by Lindholem [25].

2.4 Enumeration of actinobacterial population density

To assess the population density of actinobacteria, sediment samples were obtained using a polyvinyl corer (10 cm diameter) inserted into the sediments. Prior to sampling at each station, the corer was sterilized with alcohol. The central portion of the top 2 cm of sediment samples was extracted using a sterile scoop. These sediment samples were then transferred to sterilized polythene containers and appropriately labeled. The collected sediment samples were aseptically air-dried. After 1 week, the sediment samples were blended with an equal weight of $CaCO_3$ (1:1) and incubated at 55°C for 5 minutes [26]. The density of actinobacterial populations was determined using the spread plate method with various media viz., Kuster's agar (KUA) [27], Starch Casein Agar (SCA) [28], Actinomycetes isolation agar (AIA) [29], Starch Casein nitrate agar (SCN) [30], and AV agar (AVA) [26]. A gram of pretreated sediment sample underwent serial dilution using sterile seawater, and 0.1 ml of each serially diluted sample was introduced onto petri plates containing the respective agar medium. The samples were spread evenly using an "L" shaped glass spreader. Subsequently, the petri plates were inverted and incubated at 37°C for 7 days. Leathery colonies of actinobacteria that emerged on the petri plates were counted from the 5th day to the 28th day. The population density of actinobacteria is expressed as colony-forming units (CFU) per gram of the sample in the text. All colonies grown on the petri plates were individually streaked and sub-cultured to ensure axenic purity and were maintained on slants [26].

2.5 Statistical analysis

A graphical representation illustrating the values of water and sediment physico-chemical parameters was generated using MS Office Excel 2007 software. Statistical analysis was conducted to explore the relationship between these parameters and actinobacterial population density, utilizing Statistical Package for Social Sciences (SPSS). Principal component analysis (PCA) and cluster analysis, implemented through Plymouth Routines In Multivariate Ecological Research (PRIMER) software, were employed to identify trends among highly correlated physico-chemical parameters and actinobacterial population density.

3. Results

3.1 Physico-chemical parameters

Temperature (**Figure 2**): Air temperature varied from 29°C to 23°C, during the present period of study. The lower value (23°C) was recorded at station 5 and the higher value (29°C) was recorded at station 3. Surface water temperature varied from 26°C to 24°C, during the present study, with a lower value (24°C) at station 5 and the higher value (26°C) was recorded at stations 1, 3, and 6. Sediment temperature varied from 27°C to 23°C, registering a lower value (23°C) at station 4 and the higher value (27°C) at stations 3 and 6.

Salinity (**Figure 3**): Water salinity fluctuated from 18 to 35 psu, during the present period of study. The lower water salinity was recorded at station 4 (18 psu) and the higher value was recorded at station 6 (35 psu). Sediment salinity fluctuated from 17 to 34 psu, with the lower value (17 psu) at station 4 and the higher value (34 psu) at stations 5 and 6.

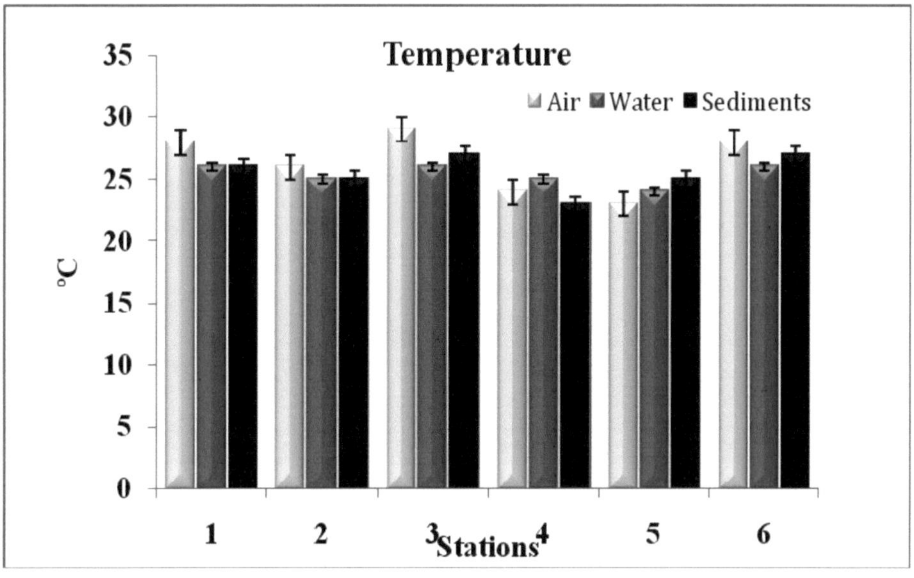

Figure 2.
Air, water and sediment temperature recorded at six stations of the Havelock Island.

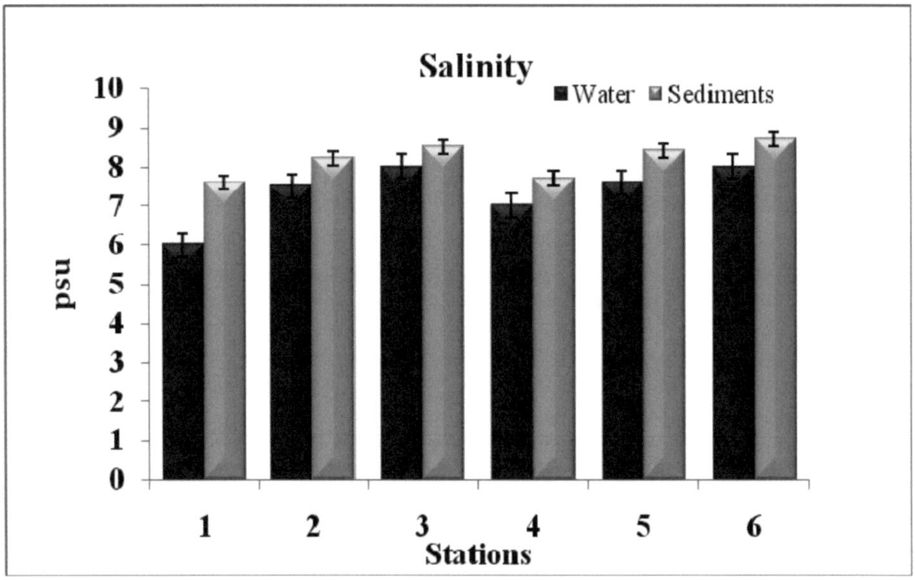

Figure 3.
Water and sediment salinity recorded at six stations of the Havelock Island.

Hydrogen-ion concentration (pH) (**Figure 4**): A narrow range of fluctuation in pH of the water was exhibited and it varied from 6 to 8. The higher pH value (8) was recorded at stations 3 and 6 and the lower pH value (6) was recorded at station 1. A narrow range of fluctuation in pH of the sediments was also recorded which varied from 7.6 to 8.7, with a higher value (8.7) at station 6 and the lower value (7.6) at station 1.

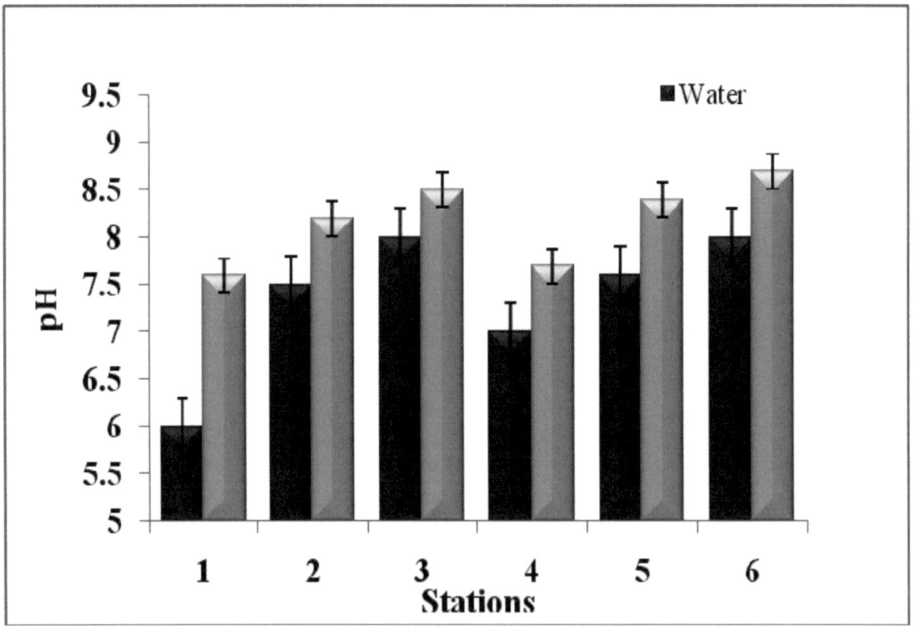

Figure 4.
Water and sediment hydrogen-ion concentration recorded at six stations of the Havelock Island.

DO concentration (**Figure 5**): DO concentration of water ranged between 2.10 and 4.60 ml l^{-1}. Lower value (2.10 ml l^{-1}) at station 1 and the higher value (4.60 ml l^{-1}) at station 5 were recorded.

EC (**Figure 6**): EC of the sediments fluctuated from 2 to 5.5 dS m^{-1}, registering the lower value (2 dS m^{-1}) at station 3 and higher value (5.5 dS m^{-1}) at station 1.

TOC (**Figure 7**): Sediment TOC content exhibited wide variations at the stations investigated, registering the lower value (0.55 mg g^{-1}) at station 6 and higher value (8 mg g^{-1}) at station 4.

Sediment nitrogen (**Figure 8**): Sediment nitrogen content varied from 2.63 to 10.88 mg g^{-1}, recording the lower value at station 2 and higher value at station 4.

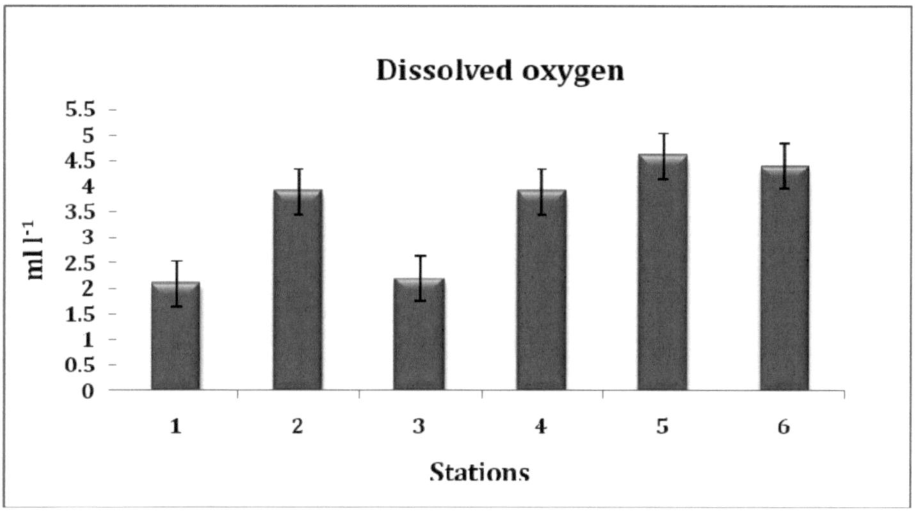

Figure 5.
Dissolved oxygen concentration recorded at six stations of the Havelock Island.

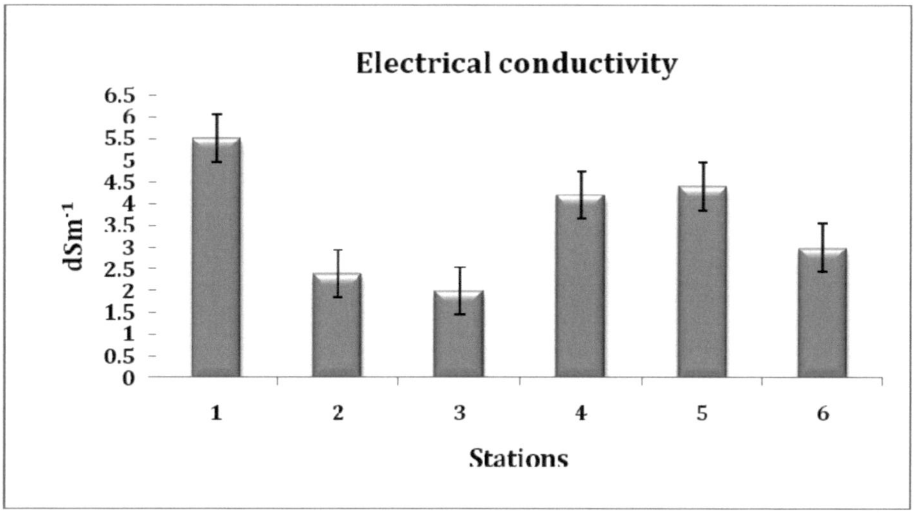

Figure 6.
Electrical conductivity recorded at six stations of the Havelock Island.

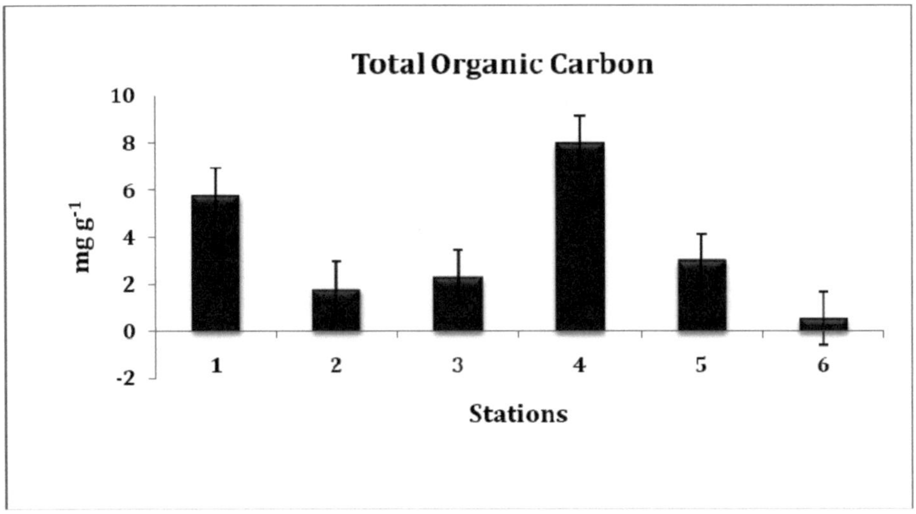

Figure 7.
Sediment total organic carbon content recorded at six stations of the Havelock Island.

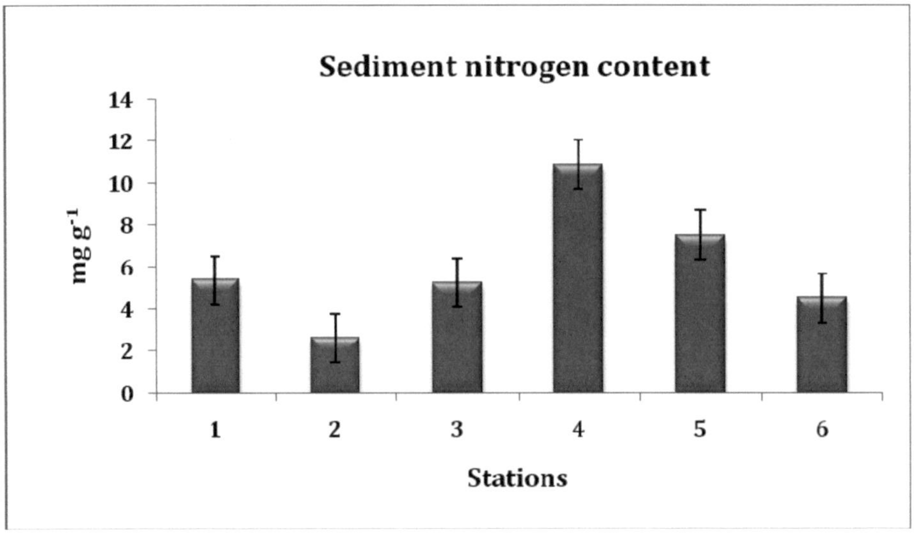

Figure 8.
Sediment nitrogen content recorded at six stations of the Havelock Island.

Sediment phosphorous (**Figure 9**): Sediment phosphorous content recorded the higher value (4.50 mg g^{-1}) at station 3 and lower value (3.13 mg g^{-1}) at station 4.

Sediment potassium (**Figure 10**): Sediment potassium content showed fluctuation, registering higher value (60.88 mg g^{-1}) at station 4 and lower value (11.88 mg g^{-1}) at station 2.

Sediment texture (**Figure 11**): Sediment texture in terms of sand, silt, and clay was in the range of 65–95%; 3.5–9.4%, and 1.2–30% respectively. Lower content of sand (65%) was recorded at station 4 and higher content (95%) was recorded at station 6. Lower silt content (3.5%) was noticed at station 1 and higher value (9.4%) was recorded at station 5. Clay content was lower (1.2%) at station 6 and higher (30%) at station 4.

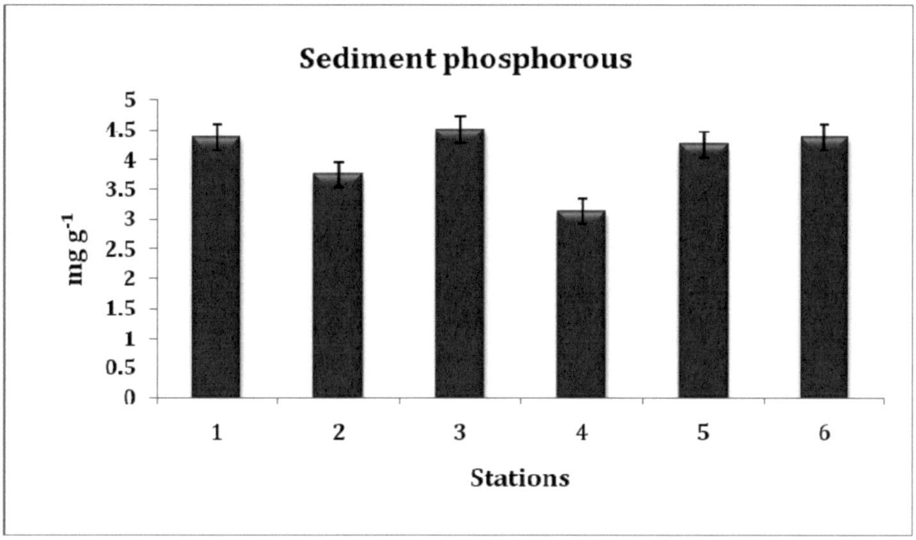

Figure 9.
Sediment phosphorous content recorded at six stations of the Havelock Island.

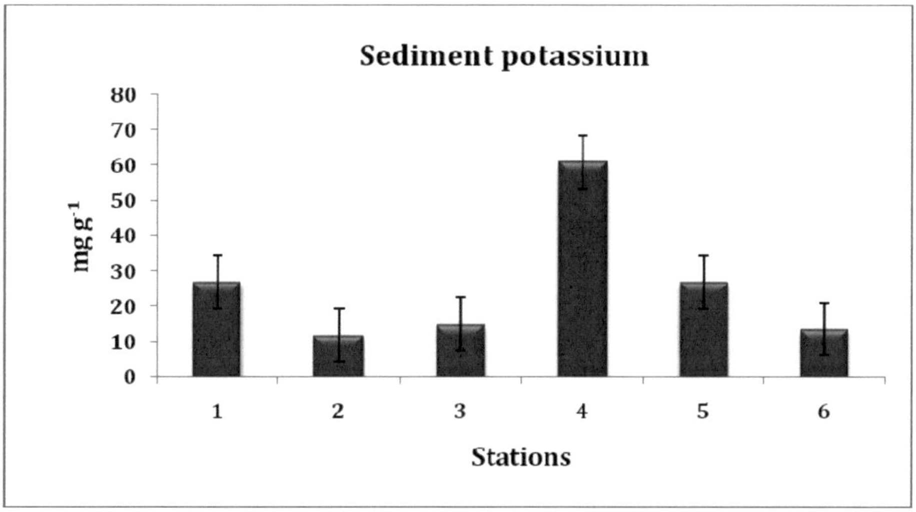

Figure 10.
Sediment potassium content recorded at six stations of the Havelock Island.

3.2 Actinobacterial population density

The actinobacterial population density (**Figure 12**), ranged from 2×10^3 to 21×10^3 CFU g^{-1}, with the highest density (21×10^3 CFU g^{-1}) observed in Kuster's agar and the lowest density (2×10^3 CFU g^{-1}) in Actinomycetes isolation agar. Within the Kuster's agar medium, the highest actinobacterial population density (21×10^3 CFU g^{-1}) was noted at station 4, while the lowest density (2×10^3 CFU g^{-1}) was observed at station 3. In Starch Casein agar medium, the highest density (17×10^3 CFU g^{-1}) occurred at station 4, and the lowest density (3×10^3 CFU g^{-1})

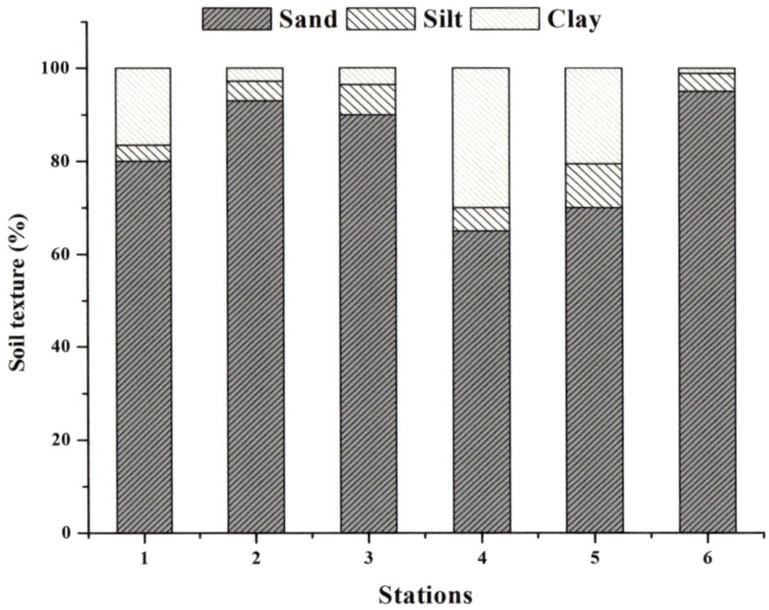

Figure 11.
Sediment texture recorded at six stations of the Havelock Island.

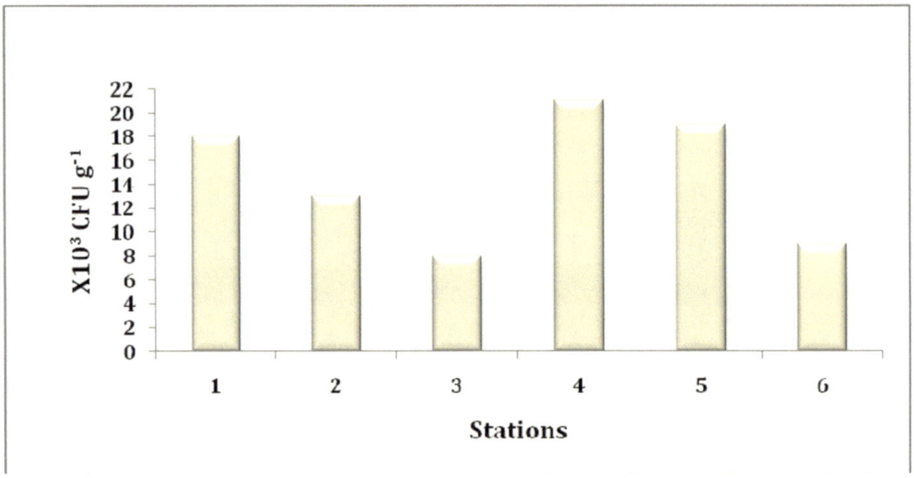

Figure 12.
Actinobacterial population density recorded at six stations of the Havelock Island.

at station 3. Actinomycetes Isolation Agar medium exhibited a higher density (9×10^3 CFU g^{-1}) at station 4 and a lower density (2×10^3 CFU g^{-1}) at station 5. Starch Casein Nitrate Agar medium displayed a higher actinobacterial population density (14×10^3 CFU g^{-1}) at station 4, while a lower density (5×10^3 CFU g^{-1}) was recorded at station 6. In the AV agar medium, a higher actinobacterial population density (11×10^3 CFU g^{-1}) was observed at station 4, with the lower density (3×10^3 CFU g^{-1}) noted at station 3.

4. Discussion

In addition to the physical, biological, and chemical factors shaping microbial niches in marine sediments, hydrodynamic forces exert a significant influence, particularly in the shallow nearshore environment. These forces are directly linked to microorganisms and the productivity of water bodies. While certain organisms exhibit adaptability across a broad range of conditions, others prove highly sensitive to environmental changes [31, 32]. So, during this direction, the present objective was to analyze the vital physico-chemical parameters (air temperature, surface water temperature, sediment temperature, water salinity, sediment salinity, water pH, sediment pH, electrical conductivity, DO, sediment nitrogen, potassium and phosphorus, TOC and soil texture) of the coastal wetland areas of the Havelock Island of the Andamans to find out their influence on the marine actinobacterial population density.

Generally, the surface water temperature will be influenced by solar radiation, evaporation, and freshwater influx [33]. Fluctuations in temperature can lead to significant changes in seawater properties, consequently affecting the life it sustains. Temperature variations contribute to alterations in water density, salinity, and dissolved oxygen levels. The growth and reproduction of aquatic life are also influenced by temperature-dependent processes [34]. Surface water temperature will be higher than that of the sediment temperature [35] because it decreases steadily from the surface to deep waters and sediments [3]. Supporting this view, in the present study, the air temperature was higher than that of the surface water and sediments in Havelock Island. This finding aligns with the research conducted by Ramaraju et al. [36], which documented a range of surface water temperatures from 27.5°C to 28.5°C in the southern Andaman Sea during winter and the study further noted a gradual decrease in temperature from the sea surface to depth. Statistically, air temperature showed a positive correlation with surface water temperature ($r = 0.944$, $P < 0.01$), sediment temperature ($r = 0.817$, $P < 0.05$), and sand ($r = 0.756$, $P < 0.05$).

Salinity is one of the most important factors that determine the composition of biological elements in the marine environment [37]. Salinity fluctuation would definitely affect the biological characteristics of the marine environment. In the present study, water salinity fluctuated from 18 to 35 psu and the salinity of the sediments fluctuated from 17 to 34 psu. Lower water and sediment salinity was found at station 4 and it could be due to the influence of freshwater prevailing in this mangrove creek area as compared to other stations. Maximum water and sediment salinity was found at station 6, which could be attributed to the high degree of evaporation and dominance of neritic waters, from the open sea [38]. Statistically, water salinity showed a positive correlation with sediment salinity ($r = 0.992$, $P < 0.01$) and sediment pH ($r = 0.796$, $P < 0.05$). Sediment salinity exhibited a positive correlation with sediment pH ($r = 0.808$, $P < 0.05$) and potassium content ($r = 0.740$, $P < 0.05$).

Water pH plays an important role by influencing all aquatic organisms, including microbes [39]. pH (water and sediments) values, ranging from 6 to 8 observed in the present study might favor the growth and survival of many types of microorganisms, as reported by Jay [40]. In the present study, sediment pH was higher (average 8.2) than that of water pH (average 7.4). In general, fluctuations in pH values are ascribed to factors, such as the removal of CO_2 through photosynthesis via bicarbonate degradation, dilution of seawater due to freshwater influx, changes in salinity and temperature, and the decomposition of organic matter [41]. In addition, pH might often change with time due to temperature-salinity changes and biological activity [42].

Statistical analysis revealed that the water pH had a significant positive correlation with sediment pH (r = 0.922, $P < 0.01$).

DO concentration in water fluctuated from 4.60 to 2.10 ml l^{-1}. Higher DO content was recorded at station 5, followed by station 6, due probably to the neritic influence. Lower dissolved oxygen concentration was recorded at station 1. This could be due to higher microbial utilization of DO. DO exhibited a negative correlation with both temperature and salinity. It is widely recognized that the dissolution of oxygen in seawater is influenced by variations in temperature and salinity [43].

EC is an indicator of higher concentrations of trace metals, which are capable of conducting electricity at much higher levels [44]. The range of EC was 1.7–6.4 dS m^{-1} in the post-tsunami period in the Andaman waters [45]. In the present study, EC of the sediments fluctuated from 2 to 5.5 dS m^{-1}. Station 4 had higher EC level (5.5 dS m^{-1}) as this station is a dense mangrove habitat with more decomposed soil and station 3 (sandy beach) had lower EC level (2 dS m^{-1}). Swarnakumar [46] reported that based on the nature of the chemical characteristics of the soil, there will be significant variations in the EC.

In the present study, TOC content closely followed the sediment type, i.e., sediments low in clay content had lower amount of TOC and as the clay content increased, TOC content also increased, as reported by Reddy and Hariharan [47]. TOC value was lower at station 6. It was higher at station 4 followed by station 1, as these stations are mangrove habitats. Therefore, higher organic carbon content in the mangroves could be due to the mangrove and terrestrial detritus present in the suspended matter [48] and a higher rate of sedimentation, in addition to the fine nature of sediments (clayey and silt sediments) [49]. Decomposition of mangrove foliage and other vegetative remains in the sediments would also contribute to higher TOC [50]. Statistically, TOC exhibited a positive significant correlation with clay (r = 0.922, $P < 0.01$) and actinobacterial population density (r = 0.785, $P < 0.05$) and a negative significant correlation with sand (r = −0.872, $P < 0.05$).

Nutrients would promote the expansion and metabolic activity of the microbial communities within the marine environment [51]. This could be true in the coastal habitats of Havelock Island where sediments of nitrogen (2.63–10.88 mg g^{-1}), phosphorus (3.13–4.50 mg g^{-1}), and potassium (11.88–60.88 mg g^{-1}) were recorded. Statistical analysis revealed that the sediment nitrogen content had a significant positive correlation with TOC (r = 0.906, $P < 0.01$), clay (r = 0.884, $P < 0.01$), and phosphorus (r = 0.777, $P < 0.05$). Phosphorus content also had a significant positive correlation with TOC (r = 0.936, $P < 0.01$) and clay (r = 0.905, $P < 0.01$).

Sediment soil texture also plays an important role in the distribution and abundance of marine microorganisms. A significant correlation between actinobacteria and clay fraction of the sediments has revealed a positive relationship between the microbes and sediments. Higher microbial population density mainly coexists with clayey sediments than the sandy sediments [52]. Similarly, in the present study, the actinobacterial population density was higher at station 4 followed by stations 5 and 1, which had more clay content. Correlation matrix also revealed that the sand content showed a negative significant correlation (r = −0.907, $P < 0.01$) whereas the clay content had a significant positive correlation (r = 0.932, $P < 0.01$). It is worth mentioning here that the clayey sediments would contain sufficient nutrients and organic matter content for the propagation of microbes, including actinobacteria [39].

In this investigation, hierarchical cluster analysis was employed to examine the similarities or differences among stations and physico-chemical parameters. The dendrogram, representing the analyzed stations and physico-chemical parameters, provides a visual overview of the clustering process (refer to **Figures 13** and **14**). The dendrogram revealed three distinct clusters. Cluster 1 aligns with station 4, Cluster 2 encompasses stations 3, 2, and 6, while Cluster 3 corresponds to stations 1 and 5 (**Figure 13**).

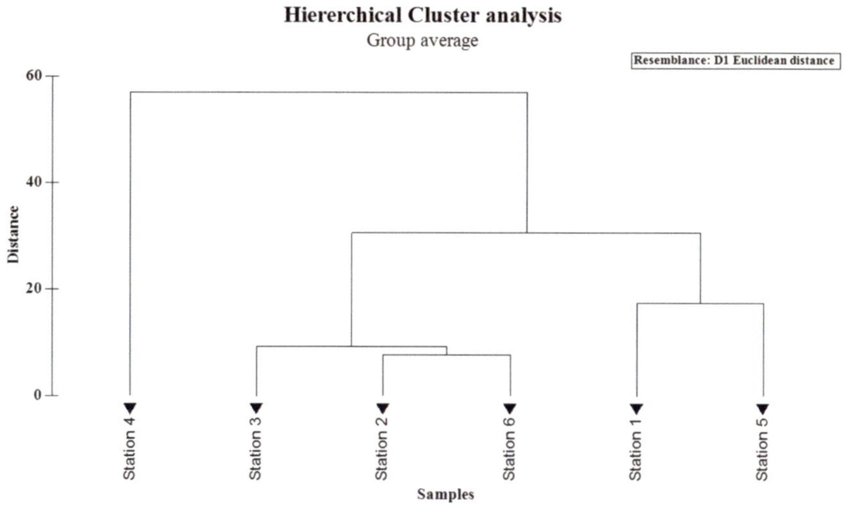

Figure 13.
Dendrogram of the cluster resemblance between physico-chemical parameters and six stations of the Havelock Island.

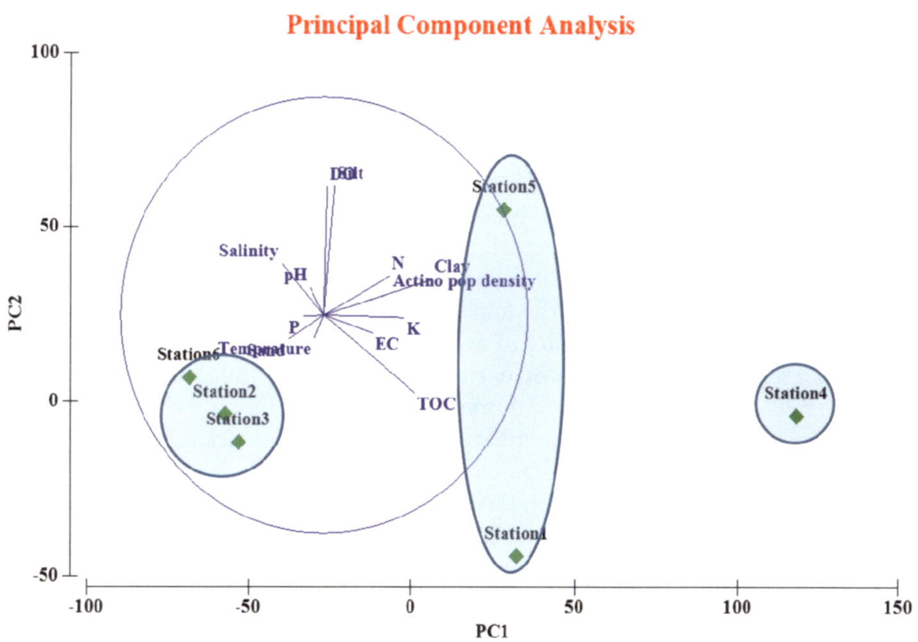

Figure 14.
Principal compound analysis plot of the physico-chemical parameters and actinobacterial population density recorded at six stations of the Havelock Island.

Cluster 1 (station 4) corresponds to a dense mangrove habitat with favorable physico-chemical conditions, supporting higher actinobacterial density, than the stations of the other clusters 2 and 3. In Cluster 2 (stations 3, 2, and 6), the station 3 clade showed a very close distance similarity with station 2 (8.929) and station 6 (9.599). These stations are coral and sandy habitats, having similar physico-chemical features. In the cluster 3 (stations 1 and 5), station 1 clade showed a close distance with station 5 (17.482), where there is the occurrence of monospecific stands of seagrasses (station 5) and station 1, occupied by patches of mangroves (station 1). Such a vegetated situation at stations 1 and 5 could be the reason for having similar physico-chemical features and actinobacterial density.

While the dendrogram can elucidate the structure of the dataset, it lacks the capacity to interpret the observed patterns in relation to the original parameters. To address this limitation, the dataset of target physico-chemical parameters underwent principal component analysis (PCA). This analytical approach was employed to ascertain which parameters significantly contribute to the variation in the actinobacterial density at the six stations of Havelock Island. PCA was utilized to discern trends among highly correlated physico-chemical parameters and microbial density, revealing a clear relationship between the stations and the physico-chemical parameters influencing actinobacterial population density.

In the PCA, a higher correlation component score was obtained between the actinobacterial population density and the physico-chemical parameters at station 4 (118), followed by station 1 (32.2) and station 5 (28.1). In the PC1 axis, these three stations had a positive significant correlation with clay, TOC, potassium (K), actinobacterial population density, nitrogen content, and EC. It could be due to the fact that these stations are occupied by mangroves (stations 4 and 1) and seagrasses (station 5). Whereas, stations 3, 2, and 6 had negative correlation component scores (−53, −57.3, and −68.2 respectively), thus proving that these three stations had a positive significant correlation with sand, phosphorus, and temperature. It could be ascribed to the fact that the two stations (stations 3 and 6) are sandy beaches and station 2 is a coral reef environment. In addition, at these three stations, very poor vegetation exists, as compared to stations 4 and 1 (mangrove habitats) and station 5 (seagrass habitat) (**Figure 14**).

Microbial populations can tolerate and grow under varying environmental conditions. During the present investigation, a higher population density of actinobacteria was recorded at station 4 in all tested media. Among these, Kuster's agar medium showed higher numbers of actinobacterial colonies at all the stations. Similarly, previous literature also reported that Kuster's agar was found to produce the desired level of actinobacterial colonies in the marine samples [26, 53–57]. Station 4, a dense mangrove habitat, had higher density of actinobacterial population since mangroves would act as a major nutrient transformation system, responsible for more microbial activity [58, 59]. Further, microbial colonies may emerge following the deposition of mangrove litter, proliferate rapidly, and attain substantial densities, actively contributing to both biomineralization and biotransformation processes of minerals [60, 61].

The density of actinobacterial populations is significantly influenced by sediment nutrients, especially total organic carbon [62] and actinobacteria, being saprophytic, thrive by relying on the availability of carbon [63]. Additionally, previous studies have also reported that the mangrove ecosystem is a potential spot to understand the actinobacterial community [56, 64–66]. Correlation matrix also showed that the actinobacterial population density had a positive significant correlation with clay ($r = 0.932$, $P < 0.01$), EC ($r = 0.811$, $P < 0.05$), nitrogen ($r = 0.809$, $P < 0.05$), P ($r = 0.699$, $P < 0.05$), and TOC ($r = 0.785$, $P < 0.05$) (**Table 2**).

	AT	SWT	ST	WS	SS	WpH	SpH	DO	EC	N	P	K	TOC	Sand	Silt	Clay	APD
AT	1																
SWT	0.944**	1															
ST	0.817*	0.646	1														
WS	0.279	0.027	0.723	1													
SS	0.214	−0.051	0.696	0.992**	1												
WpH	0.054	−0.097	0.339	0.541	0.561	1											
SpH	0.21	0.018	0.608	0.796*	0.808*	0.922**	1										
DO	−0.682	−0.657	−0.368	0.242	0.254	0.44	0.379	1									
EC	−0.356	−0.195	−0.359	−0.39	−0.37	−0.824*	−0.692	−0.065	1								
N	−0.372	−0.132	−0.721	−0.942**	−.0921**	−0.735*	−0.898**	−0.246	0.665	1							
P	−0.603	−0.38	−0.692	−0.761*	−0.699	−0.228	−0.433	0.176	0.477	0.777*	1						
K	0.606	0.408	0.920**	0.729	0.740*	0.194	0.514	−0.353	−0.091	−0.61	−0.56	1					
TOC	−0.575	−0.307	−0.825*	−0.906**	−0.875*	−0.436	−0.667	0.113	0.53	0.906**	0.936**	−0.749*	1				
Sand	0.759*	0.597	0.771*	0.681	0.606	0.439	0.58	−0.167	−0.681	−0.810*	−0.906**	0.516	−0.872*	1			
Silt	−0.55	−0.701	−0.124	0.177	0.295	0.378	0.336	0.29	0	−0.088	0.34	0.147	0.06	−0.445	1		
Clay	−0.707	−0.505	−0.801*	−0.763*	−0.705	−0.541	−0.684	0.122	0.728	0.884**	0.905**	−0.58	0.922**	−0.986**	0.288	1	
APD	−.760*	−0.598	−0.820*	−0.627	−0.583	−0.671	−0.752*	0.185	0.811*	0.809*	0.699	−0.57	0.785*	−0.907**	0.2	0.932**	1

*Correlation is significant at 5% level ($P < 0.05$).
**Correlation is significant at 1% level ($P < 0.01$).
Air temperature (AT), surface water temperature (SWT), sediment temperature (ST), water salinity (WS), sediment salinity (SS), water pH (WpH), sediment pH (SpH), dissolved oxygen (DO), electrical conductivity (EC), nitrogen (N), phosphorus (P), potassium (K), total organic carbon (TOC), actinobacterial population density (APD).

Table 2.
Pearson correlations coefficient (r) values of the actinobacterial population and physico-chemical characteristics of Havelock Island.

At stations 5 and 1, there was a moderate population density of actinobacteria correlating with the occurrence of monospecific stands of seagrasses (station 5) and patches of mangroves (station 1), as revealed by the average cluster analysis, where both the stations fell under the same cluster (cluster 3). Next to the mangrove and seagrass habitats, station 2 (coral reef environment) recorded a fair actinobacterial population density. It could be possible because corals can secrete mucus, which makes up the surface mucopolysaccharide layer, a medium high in organic compounds, which constantly traps microbes of all varieties from seawater and develops a higher concentration of the community of bacteria than that of ambient seawater [67]. Stations 3 and 6, sandy beaches, with organic carbon content, 50–66.7% (lower than that of the muddy sediments), showed only a lower density of actinobacteria, compared to other habitats. In Hierarchical cluster analysis also, these three stations (3, 2, and 6) fell under the same cluster (cluster 2) with close resemblance distance.

5. Conclusion

In conclusion, the present study has thus found that the physico-chemical parameters (air, surface water, and sediment temperatures, water and sediment salinity, water and sediment pH, electrical conductivity, dissolved oxygen, sediment nutrients, total organic carbon and soil texture) have influenced the actinobacterial population density directly or indirectly, in the different coastal habitats of the Havelock Island. The mangrove environment has favored higher density of actinobacteria than the other habitats. However, it should be noted that these environmental factors and their influence are specific and applicable only to the microbial populations of the study areas of the Havelock Island and should not be generalized. Additionally, information pertaining to the water quality parameters would help understand the present status of the coast of Havelock Island, which is now open for tourism. Hence, further in-depth studies comparing our study results will help understand the current stations of hydrographical features and actinobacterial density and diversity due to the latter's ecosystems and biotechnology potentials.

Acknowledgements

The authors express their gratitude to the Dean of the Faculty of Marine Sciences and the authorities of Annamalai University for generously providing the necessary facilities. Additionally, the Author (R.G) extends thanks to the Director and Joint Directors of SCERT, Chennai, as well as Dr. P. Natarajan, Principal of the District Institute of Education and Training, Pudukkottai, for their valuable support. It is important to note that the contents and views presented in this manuscript are those of the individual author and do not necessarily reflect the perspectives or positions of the respective institutions to which the author is affiliated.

Credit authorship contribution

RG & KS: conceptualization, investigation, data curation, writing, original draft preparation, reviewing, and editing.

Declaration of competing interest

The authors affirm that they do not have any known competing interests that might have influenced the work reported in this article.

Data availability

The datasets utilized and examined in the present study can be obtained from the corresponding author upon reasonable request.

Additional information

This work is an extended version of the doctoral thesis by the same author [R. Gobalakrishnan] and the title is Ecology, diversity and bioelectricity potential of Marine actinobacteria from The Havelock Island of the Andaman India. 2013, Unpublished doctoral thesis, Annamalai University, Available from: https://www.inflibnet.ac.in. And also, authors have full copyrights to use all the figures in this chapter.

Author details

Rajagopal Gobalakrishnan[1*] and Kannan Sivakumar[2]

1 Lecturer in Botany, District Institute of Education and Training, Pudukkottai, Tamil Nadu, India

2 Centre of Advance Study in Marine Biology, Faculty of Marine Sciences, Annamalai University, Parangipettai, Tamil Nadu, India

*Address all correspondence to: gobalaldy@gmail.com

References

[1] Bakri DA, Kittaneh W. Physicochemical characteristics and pollution indicators in the intertidal zone of Kuwait: Implications for benthic ecology. Environmental Management. 1998;**22**:415-424. DOI: 10.1007/s002679900116

[2] Cevera JH, Karl D, Buckley M. Marine Microbial Diversity: The Key to Earth's Habitability. San Francisco, California: American Academic of Microbiology Colloquium Report; 2005. pp. 1-22

[3] Das S, Lyla PS, Ajmalkhan S. Spatial variation of aerobic culturable heterotrophic bacterial population in sediments of the continental slope of western Bay of Bengal. Indian Journal of Geo-Marine Science. 2007;**36**:51-58. Available from: http://nopr.niscair.res.in/handle/123456789/14

[4] Azam F, Fenchel T, Field JG, Gray JS, Meyer-Reil LA, Thingstad F. The ecological role of water column microbes in the sea. Marine Ecology Progress Series. 1983;**10**(3):257-263. DOI: 10.3354/meps010257

[5] Ducklow HW, Hill S. The growth of heterotrophic bacteria in the surface waters of warm core rings. Limnology and Oceanography. 1985;**30**:239-259. DOI: 10.4319/lo.1985.30.2.0239

[6] Alavandi SV. Relationship between heterotrophic bacteria and suspended particulate matter in the Arabian Sea. Indian Journal of Microbiology. 1990;**30**(1):89-92. Available from: https://krishi.icar.gov.in/jspui/bitstream/123456789/ 1155/1/05-1990%20Alavandi%20IJM.pdf

[7] Dong H, Jiang H, Yu B, Liu X. Impacts of environmental change and human activity on microbial ecosystems on the Tibetan Plateau, NW China. GSA Today. 2010;**20**:4-10. DOI: 10.1130/GSATG75A.1

[8] Duran R, Ranchou-Peyruse M, Menuet V, Monperrus M, Bareille G, Goni MS, et al. Mercury methylation by a microbial community from sediments of the Adour Estuary (Bay of Biscay, France). Environmental Pollution. 2008;**156**(3):951-958. DOI: 10.1016/j.envpol.2008.05.007

[9] Thiyagarajan V, Lau SCK, Mandy T, Zhang W, Qian PY. Monitoring bacterial biodiversity in surface sediment using terminal restriction fragment length polymorphism analysis (T-RFLP): Application to coastal environment. In: Ishimatsu A, Lie HJ, editors. Coastal Environmental and Ecosystem Issues of the East China Sea. Nagasaki: Terrapub, Nagasaki University; 2010. pp. 151-163

[10] Stevens H, Brinkhoff T, Rink B, Vollmers J, Simon M. Diversity and abundance of Gram positive bacteria in a tidal flat ecosystem. Environmental Microbiology. 2007;**9**:1810-1822. DOI: 10.1111/j.1462-2920.2007.01302.x

[11] Mincer TJ, Jensen PR, Kauffman CA, Fenical W. Widespread and persistent populations of a major new marine actinomycete taxon in ocean sediments. Applied and Environmental Microbiology. 2002;**68**(10):5005-5011. DOI: 10.1128/AEM.68.10.5005-5011.2002

[12] Maldonado LA, Stach JEM, Pathom-aree W, Ward AC, Bull AT, Goodfellow M. Diversity of cultivable actinobacteria in geographically widespread marine sediments. Antonie Van Leeuwenhoek. 2005;**87**:11-18. DOI: 10.1007/s10482-004-6525-0

[13] Pathom-aree W, Stach JEM, Ward AC, Horikoshi K, Bull AT, Goodfellow M. Diversity of actinomycetes isolated from Challenger Deep sediment (10,898 m) from the Mariana Trench. Extremophiles. 2006;**10**:181-189. DOI: 10.1007/s00792-005-0482-z

[14] Urakawa H, Tsukamoto KK, Ohwada K. Microbial diversity in marine sediments from Sagami Bay and Tokyo Bay, Japan, as determined by 16S rRNA gene analysis. Microbiology. 1999;**145**:3305-3315. DOI: 10.1099/00221287-145-11-3305

[15] Bull AT, Stach JEM, Ward AC, Goodfellow M. Marine actinobacteria: Perspectives, challenges, future directions. Antonie Van Leeuwenhoek. 2005;**87**:65-79. DOI: 10.1007/s10482-004-6562-8

[16] David S. Terrestrial ecosystem and carbon cycle. Global Change Biology. 2006;**1**(1):77-91. DOI: 10.1111/j.1365-2486.1995.tb00008.x

[17] Ramesh S, William A. Marine actinomycetes: An ongoing source of novel bioactive metabolites. Microbiological Research. 2012;**167**(10):571-580. DOI: 10.1016/j.micres.2012.06.005

[18] Sankar R, Ramkumar L, Rajkumar M, Sun J, Ananthan G. Seasonal variations in physico-chemical parameters and heavy metals in water and sediments of Uppanar estuary Nagapattinam India. Journal of Environmental Biology. 2010;**31**(5):681-686. Available from: https://pubmed.ncbi.nlm.nih.gov/21387922/

[19] Jayaraj KA, Jayalakshmi KV, Saraladevi K. Influence of environmental properties on macrobenthos in the Northwest Indian shelf. Environmental Monitoring and Assessment. 2007;**127**:459-475. DOI: 10.1007/s10661-006-9295-5

[20] Reddi KR, Jayaraju N, Suriyakumar I, Sreenivas K. Tidal fluctuation in relation to certain physio-chemical parameters in Swarnamukhi river estuary, East coast of India. Indian Journal of Geo-Marine Sciences. 1993;**22**:232-234. Available from: http://nopr.niscair.res.in/handle/123456789/37883

[21] Strickland JDH, Parsons TR. A Practical Handbook of Seawater Analysis, Bulletin. Vol. 167. Ottawa: Fisheries Research Board of Canada; 1972

[22] Subbiah BV, Asija GL. A rapid procedure for estimation of available nitrogen in soils. Current Science. 1956;**25**:259-260

[23] Olsen SR, Cole CV, Watanabe FS. Estimation of Available Phosphorus in Soil by Extraction with Sodium Bicarbonate, Circular, No. 939. Washington, DC: United States Department of Agriculture; 1954

[24] El-Wakeel SK, Riley JP. The determination of organic carbon in marine mud. ICES Journal of Marine Science. 1956;**22**(2):180-183. DOI: 10.1093/icesjms/22.2.180

[25] Lindholem RC. A Practical Approach to Sedimentology. Netherlands: Springer; 1987. DOI: 10.1007/978-94-011-7683-5

[26] Gobalakrishnan R. Phylogenetic diversity of culturable marine actinobacteria isolated from the Havelock Island, the Andamans, India. Ecological Genetics and Genomics. 2023;**100123**:1-8. DOI: 10.1016/j.egg.2022.100123

[27] Gobalakrishnan R, Sivakumar L. Systematic characterization of potential cellulolytic marine actinobacteria

Actinoalloteichus sp. MHA15. Biotechnology Reports. 2017;**13**:30-36. DOI: 10.1016/j.btre.2016.12.003

[28] Anupama S, Aishwarya T, Anupa B, Muskan S, Prativa S, Sagar A. Isolation, characterization and screening of antimicrobial producing actinomycetes from soil samples. International Journal of Microbiology. 2020;**2020**:1-7. Article ID: 2716584. DOI: 10.1155/2020/2716584

[29] Abdenbi El K, Soumia AA, Mohammed EH. Isolation and screening of actinomycetes producing antimicrobial substances from an extreme Moroccan biotope. The Pan African Medical Journal. 2019;**33**(329):1-9. DOI: 10.11604/pamj.2019.33.329.19018

[30] Waksman SA. Strain specificity and production of antibiotic substances. Characterization and classification of species within the Streptomyces griseus group. Proceedings of the National Academy of Sciences of the United States of America. 1959;**45**(7):1043-1047. DOI: 10.1073/pnas.45.7.1043

[31] Rusch A, Huettel M, Reimers CE, Taghon GL, Fuller CM. Activity and distribution of bacterial populations in Middle Atlantic Bight shelf sands. FEMS Microbiology Ecology. 2003;**44**:89-100. DOI: 10.1111/j.1574-6941.2003.tb01093.x

[32] Raibole M, Singh YP. Impact of physico-chemical parameters on microbial diversity: Seasonal study. Current World Environment: An International Research Journal of Environmental Science. 2011;**6**:71-76. DOI: http://dx.doi.org/10.12944/CWE.6.1.09

[33] Ajithkumar TT, Thangaradjou T, Kannan L. Physico-chemical and biological properties of the Muthupettai mangrove in Tamil Nadu. The Marine Biological Association of India. 2006;**48**:131-138

[34] Davis RA. Principles of Oceanograph. 2nd ed. New York: Addison - Wesley Publishing; 1977

[35] Muraleedharan H, Abhilash R. Physico-chemical parameters and planktons analysis of sea water of Thondi of Palk Bay, Tamilnadu. Journal of Biological Science Research. 2010;**1**(1):20-24

[36] Ramaraju DV, Gouveia AD, Murthy CS. Some physical characteristic of Andaman Sea waters during winter. Indian Journal of Geo-Marine Science. 1981;**10**:211-218. Available from: http://nopr.niscair.res.in/handle/123456789/39079

[37] Manivasagan P. Ecology, taxonomy and biological potential (antibiotic, pigment and enzyme) of marine actinobacteria from the Tamilnadu part of Bay of Bengal, India [Ph.D. thesis]. India: Annamalai University; 2009

[38] Senthilkumar S, Santhanam P, Perumal P. Diversity of phytoplankton in Vellar estuary, Southeast coast of India. In: Ayyappan S, Jena JK, Mohan Joseph M, editors. AAFSIB (Asian Fisheries Society Indian Branch), Mangalore and AoA (Association of Aquaculturists), Bhubaneswar, India. 2002. pp. 245-248

[39] Gnanam S, Sivakumar K, Vijayalakshmi S, Balasubramaniyan T. Ambience of physico-chemical characteristics to actinobacterial density in the Chennai coast of the Bay of Bengal, India. Indian Journal of Geo-Marine Science. 2013;**42**:254-265. Available from: http://nopr.niscair.res.in/handle/123456789/16948

[40] Jay JM. Modern Food Microbiology. 5th ed. New York: Van Norstrand Reinhold; 1996

[41] Rajasegar M. Physico-chemical characteristics of the Vellar estuary in relation to shrimp farming. Journal of Environmental Biology. 2003;**24**(1):95-101

[42] Srilatha G, Thilagavathi B, Varadharajan D. Studies on the physico-chemical status of Muthupettai mangrove, South east coast of India. Advances in Applied Science Research. 2012;**3**(1):201-207

[43] Vijayakumar SK, Rajesh KM, Mendon MR, Hariharan V. Seasonal distribution and behaviour of nutrients with reference to tidal rhythm in the Mulki estuary, southwest coast of India. Journal of the Marine Biological Association of India. 2000;**42**(1&2):21-31. Available from: http://mbai.org.in/uploads1/manuscripts/Chpater_02%20(21-31)1856714598.pdf

[44] Chaudhuri GS, Dinesh R, Sheeja TE, Raja R, Jeykumar V, Srivastava RC. Physico-chemical, biochemical and microbial characteristics of soils of mangroves of the Andamans: A post-tsunami analysis. Current Science. 2009;**97**:98-102. Available from: https://www.jstor.org/stable/24112089

[45] Nayak AK, Damodaran T, Singh CS, Jha SK, Raja D, Mishra VK, et al. Post tsunami changes in soil properties of Andaman islands India. Environmental Monitoring and Assessment. 2010;**170**:185-193. DOI: 10.1007/s10661-009-1225-x

[46] Swarnakumar NS. Studies on the actinobacteria of the coastal environment of Nicobar group of islands with special reference to anticancer properties [Ph.D. thesis]. India: Annamalai University; 2010

[47] Reddy HRV, Hariharan V. Distribution of nutrients in the sediments of the Netravathi-Gurupur estuary, Mangalore. Indian Journal of Fisheries. 1986;**33**(1):123-126

[48] Jagtap TG. Seasonal distribution of organic matter in mangrove environment of Goa. Indian Journal of Geo-Marine Science. 1987;**16**:103-106

[49] Raghunath R, Sreedhara Murthy TR. Carbonate and organic matter studies of the shelf of Kasargod, west coast of India. Indian Journal of Geo-Marine Science. 1996;**25**:355-357. Available from: http://nopr.niscair.res.in/handle/123456789/37152

[50] Ramanathan AL. Sediment characteristics of the Pichavaram mangrove environment, Southeast coast of India. Indian Journal of Geo-Marine Science. 1997;**26**:319-322. Available from: http://nopr.niscair.res.in/handle/123456789/36329

[51] Trupti KV, Dave BP. Effect of addition of nitrogen, phosphorus and potassium fertilizers on biodegradation of crude oil by marine bacteria. Indian Journal of Geo-Marine Science. 2010;**39**:143-150. Available from: http://nopr.niscair.res.in/handle/123456789/8558

[52] Nair SL, Bharathi PA, Achuthankutty CT. Distribution of heterotrophic bacteria in marine sediments. Indian Journal of Geo-Marine Science. 1978;**7**:18-22

[53] Saadoun I, Al-momani F. Bacterial and Streptomyces flora of some Jordan valley soils. Actinomycetes. 1996;**7**:95-99. Available from: http://hdl.handle.net/1807/22671

[54] Raghavendrudu G, Kondalarao B. Studies on the distribution of actinobacteria in the Gaderu mangroves of Gautami Godavari estuarine system,

East coast of India. Journal of the Marine Biological Association of India. 2007;**49**:246-249. Available from: http://mbai.org.in/php/journaldload.php?id=2041&bkid=93

[55] Baskaran R, Vijayakumar R, Mohan PM. Enrichment method for the isolation of bioactive actinomycetes from mangrove sediments of Andaman islands, India. Malaysian Society for Microbiology. 2011;7:26-32

[56] Sethubathi GVB, Sivakumar K, Thangaradjou T, Kannan L. Ecology and population density of the marine actinobacteria of little Andaman island India. Indian Journal of Geo-Marine Science. 2013;**42**:390-401. Available from: http://nopr.niscair.res.in/handle/123456789/19653

[57] Gobalakrishnan R, Bhuvaneswari R. Microbial fuel cells potential of marine actinobacteria Actinoalloteichus sp. MHA15 from the Havelock Island of the Andamans, India. Biotechnology Research and Innovation. 2019;**3**(1):144-158. DOI: 10.1016/j.biori.2019.01.003

[58] Alongi DM, Christoffersen P, Tirendi F. The influence of forest type on microbial nutrient relationships in tropical mangrove sediments. Journal of Experimental Marine Biology and Ecology. 1993;**171**(2):201-223. DOI: 10.1016/0022-0981(93)90004-8

[59] Holguin G, Vazquez P, Bashan Y. The role of sediment microorganisms in the productivity, conservation, and rehabilitation of mangrove ecosystems: An overview. Biology and Fertility of Soils. 2001;**33**:265-278. DOI: 10.1007/s003740000319

[60] Kathiresan K. Degradation and destruction of mangroves. In: Kathiresan K, Ajmal Khan S, editors. International Training Course on Coastal Biodiversity in Mangroves Ecosystems. India: UNU-INWEH-UNESCO; 2007. pp. 476-483

[61] Gonzalez-Acosta B, Bashan Y, Hernandez-Saavedra NY, Ascenaio F, Cruz Aguero G. Seasonal seawater temperature as the major determinant for populations of culturable bacteria in the soils of an in-tact mangrove in an Arid region. FEMS Microbiology Ecology. 2006;**55**:311-321. DOI: 10.1111/j.1574-6941.2005.00019.x

[62] Das S, Lyla P, Ajmalkhan S. Distribution and generic composition of culturable marine actinomycetes from the marine sediments of continental slope of Bay of Bengal. Chinese Journal of Oceanology and Limnology. 2008;**26**:166-177. DOI: 10.1007/s00343-008-0166-5

[63] Promod KC, Dhevendaran K. Studies on phosphobacteria in Cochin backwater. Journal of the Marine Biological Association of India. 1987;**29**(1&2):297-305

[64] Karthikeyan P, Senthilkumar G, Panneerselvam A. Actinobacterial diversity of mangrove environment of the Palaverkadu mangroves, East Coast of Tamilnadu, India. International Journal of Current Microbiology and Applied Sciences. 2014;**3**:145-154. Available from: http://www.ijcmas.com/vol-3-1/P.Karthikeyan%20G.Senthilkumar%20and%20A.Panneerselvam.pdf

[65] Mangamuri UK, Vijayalakshmi M, Sudhakar P. Exploration of actinobacteria from mangrove ecosystems of Nizampatnam and Coringa for antimicrobial compounds and industrial enzymes. Biotechnology Journal International. 2014;**4**:173-184. DOI: 10.9734/BBJ/2014/6424

[66] Poongodi S, Sivakumar K, Sivasankar P, Kannan L. Ecological

factors and their influence on actinobacterial density from different coastal habitats of Neil island, Andamans. Indian Journal of Geo-Marine Sciences. 2016;**45**(07):911-919. Available from: http://nopr.niscair.res.in/handle/123456789/35137

[67] Ritchie KB. Regulation of microbial population by coral surface mucus and mucus-associated bacteria. Marine Ecology Progress Series. 2006;**322**:1-14. DOI: 10.3354/meps322001

Chapter 6

Cyanobacterial Mats in Stradline Holocene Carbonate: Evaporite Deposits along the Coast of Southern Arabian Gulf – Potential Sources of Petroleum in Ancient Carbonate Reservoirs

Abdulrahman Alsharhan and Christopher Kendall

Abstract

Lining the sheltered inner lagoon behind the barrier islands in the southern Arabian Gulf in the United Arab Emirates coastline are a series of algal flats that are underlain by some 30 cm of compacted, organic, peat-like material derived from the accumulation of the remains of living algae. This algal mat is prograding seaward and is being covered by supratidal wash-over carbonates and displacive evaporites. Both the organic material and interbedded carbonates and evaporites were examined to determine their distribution, petrography, and chemical composition. The chemical analysis of organics will involve the use of GC/MS and Rock-Eval pyrolysis techniques. TOC of these sediments has a wide range of organic enrichment (0.46–8.40%), and whole rock pyrolysis yields moderately high hydrocarbon indices (HI) 389–597, typical of marine Type II kerogens. This association of high concentrations of organic matter interbedded with carbonates and evaporites is unique to the Holocene but is believed to have occurred in the geological past, where such associations may represent potential source rocks for ancient carbonate petroleum reservoirs in some oil fields in the Middle East, United States, and Canada.

Keywords: Holocene cyanobacterial mat, potential source of hydrocarbon, Abu Dhabi, UAE, petroleum

1. Introduction

The coastal flats of Abu Dhabi are part of the southeastern Arabian Gulf carbonate province and are characterized by a series of carbonate barrier islands and associated shallow lagoons. The southern coast of these lagoons is flanked by long cyanobacterial flats that pass landward into supratidal salt flats or sabkhas (**Figure 1**). Observations and sample collections were made along a road just to the south of Al Qanatir barrier

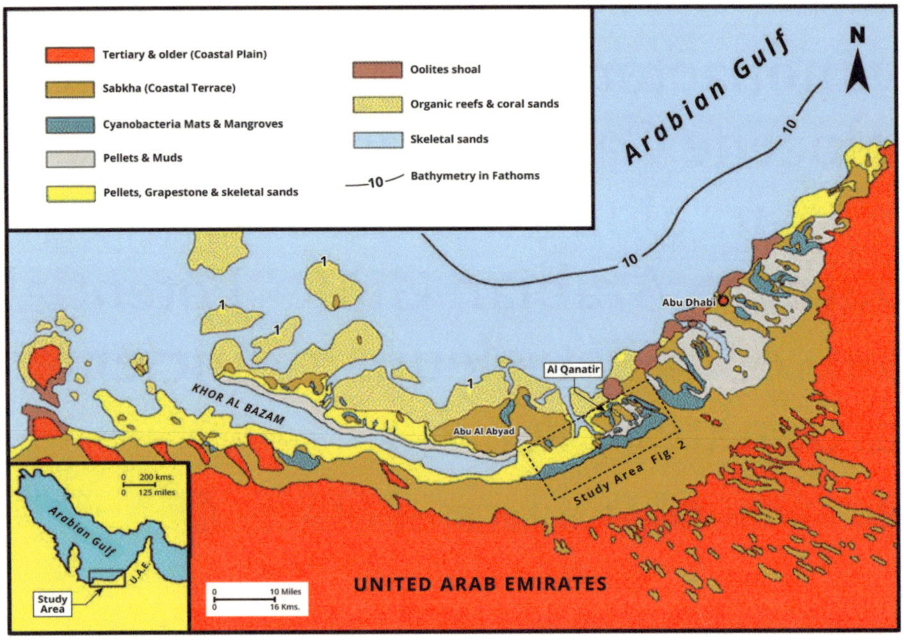

Figure 1.
Map of the general marine sedimentary facies of the coast of Abu Dhabi, United Arab Emirates [1].

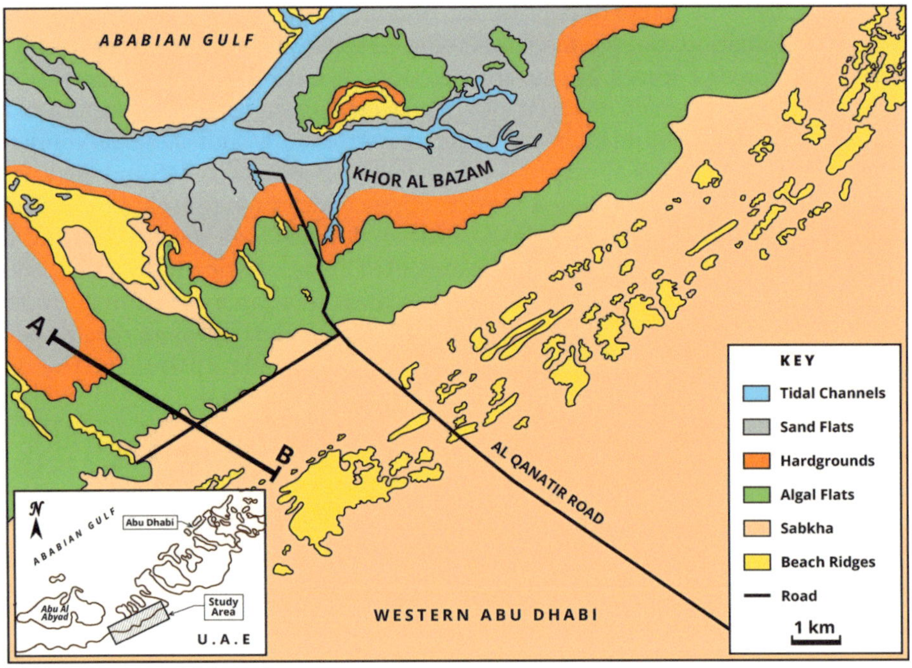

Figure 2.
*Coastal sabkha and tidal flats in the vicinity of Al Qanatir island. Note the east-west road at the top of the tidal flat acts as a barrier of floodwaters. Line A-B shows the location of transect cores in **Figure 3** [2].*

island where our study area is located (**Figure 2**) to determine the surface and subsurface relationships between the organic and inorganic facies. The area of concentration lies between the subtidal and supratidal exposures of a foraminiferal-gastropod sand/mud, with the supratidal exposure represented by an ancient and now partially lithified beach ridge that topographically subdivides the lower and upper sabkha, outlining the upper extent of storm tide, and the subtidal exposure representing present lagoon carbonate sand and mud deposition.

The Abu Dhabi carbonate-evaporite complex has been the focus of significant geological research. However, few of these studies have generated the information necessary to establish the source rock potential of the cyanobacterial remains [3–5]. The southern shore of the Khor Al Bazam coastal lagoon of Abu Dhabi is lined by a wide, long, algal tidal flat (**Figure 1**). At the east end of the lagoon, the largest algal flat parallels the coast for 42 km and to the west a smaller one parallels the coast for some 9 km. These flats, part of the seaward edge of the prograding coastal plain, have an average width of about 2 km and a thickness of between 5 and 30 cm. In some areas, they extend landward in the subsurface for more than 2 km beneath the thin veneer of evaporites and wind-blown and storm-wash over sediment [6, 7]. Using carbon dating, it has been determined that the algal mat has prograded 7 km in the last 4000 years [8].

The study area is located on the tidal flat south and southeast of Al Qanatir and east of the Khor Al Bazam (**Figure 1**), and there is access to the flats via a road running perpendicular to the Tariff-Abu Dhabi highway toward Al Qanatir. Traverse location was measured from the road and by using conventional siting methods with a Brunton compass on objects located on satellite image map.

Traverse is made from earlier Holocene beach ridges to the present shoreline (**Figure 3**). Traced from landward to seaward, the surface sediment types on this traverse vary and reflect the local topography and a relationship to the tide range or where the sabkha slope adjacent to beach ridges is disrupted by tidal channel

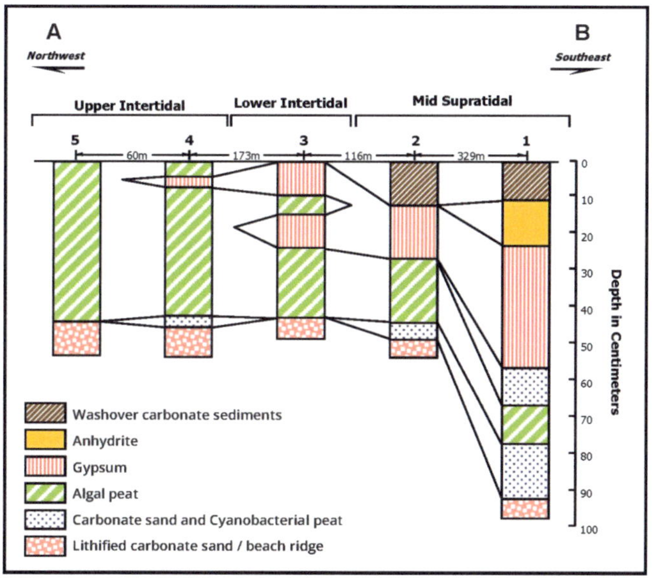

Figure 3.
Interpretation of transect cores showing the sedimentary facies. Location of the section on ***Figure 2****.*

depressions and by elevated hardground tepee structures [9, 10]. From field observation further inland, the algal mat becomes more deeply buried, and gross morphologic structures are less well preserved. Also, from field observations below the gypsum mush, polygonal and mammillated mats are easily distinguishable in buried sections. Further inland and beneath the supratidal facies, only the polygonal mat was identified [1], and crenulated algal mat facies was not evident in the subsurface.

Several major oil fields in the world occur in ancient carbonate/evaporite sequences similar to those of the study and could also have contained similar algal deposits. The Abu Dhabi carbonate-evaporite complex has been the focus of significant geological research. However, few of these studies have generated the information necessary to establish the source rock potential of the cyanobacterial remains.

2. Objective and scope of study

The primary objective of the study is to develop an understanding of the petroleum source rock potential of algal deposits forming in association with tidal flat carbonates on the coast of Abu Dhabi (UAE). We collected both the organic material and the interbedded carbonates and evaporites to determine their distribution, petrography, and chemical composition. Chemical analysis of the organics was performed with the use of combined gas chromatography/mass spectroscopic (GC/MS) and GC/FTIR/FID techniques, as well as other conventional petroleum exploration techniques (such as Rock-Eval pyrolysis, carbon isotopic analysis, and biomarker fingerprints).

3. Methods

A field investigation was made between 2002 and 2006 in the Khor Al Bazam just west of the Jazirat Al Qanatir road (**Figure 2**). Samples of cyanobacterial material and associated carbonate sediments were collected along six transects perpendicular and longitudinal to the shoreline. Only one general traverse presented here and at selected locations of trenches was dug, in which eight vertical 3-inch diameter cores of sediment were collected. In order to investigate postburial effects, some of these cores were taken in areas where the algal mat extends inland beneath other sediments. All core and surface samples were sealed in plastic containers to ensure preservation. These samples collected from cores in **Figure 3** were investigated in terms of their: (A) surface conditions at deposition; (B) geometry and relationship to surrounding facies; and (C) organic chemical composition from whole rock, bitumen and kerogen analysis [2].

Representative samples of algal material were analyzed microscopically for petrographic and mineralogic components. The kinds of organic components present were established as well as their degree of alteration, and their relationship to mineralogic components. Standard petrographic thin sections of concentrates were also prepared and analyzed in both white and fluorescent light.

For standard organic carbon analysis total organic carbon (TOC), a LECO CS-344 carbon and sulfur analyzer was utilized. Whole rock pyrolysis was accomplished by analyzing dried, ground sediments with a Delsi Rock-Eval II pyrolysis unit. Solid organic material (kerogen precursor) was isolated by removing the mineral matrix using alternating hydrochloric acid and water washes. Hydrofluoric acid digestion was utilized to remove the small number of remaining silicates.

Elemental composition (C, H, N, O, S) of the solid organic material was determined by combustion analysis utilizing a Carlo Erba model 1106 elemental analyzer. Ash content was determined by calculating the difference between the sample and the total C, H, N, O, S content. Then organic elemental composition was calculated on an ash-free basis. Carbon isotope measurements of isolated solid organic material were accomplished by combusting the organic matter and cryogenically trapping carbon dioxide. Then the carbon dioxide gas was introduced into a Finnigan Delta E isotope ratio mass spectrometer. Samples were subjected to pyrolysis gas chromatography/mass spectroscopic (GC/MS) analysis using a Finnigan 4021 mass spectrometer equipped with a Finnigan 4521 El source.

4. Field observations

4.1 Carbonate-evaporites sedimentary facies

The main sedimentary facies found between the barrier islands to beach ridges and sand flats in the study area is shown from Landsat image in **Figure 4** [11]. From the ancient beach sediment to the present shoreline, the land cover varies from: thin (3 mm) halite, polygonally cracked surface; convoluted halite crust; moist carbonate mud-gypsum mush; and series of distinct algal mat types.

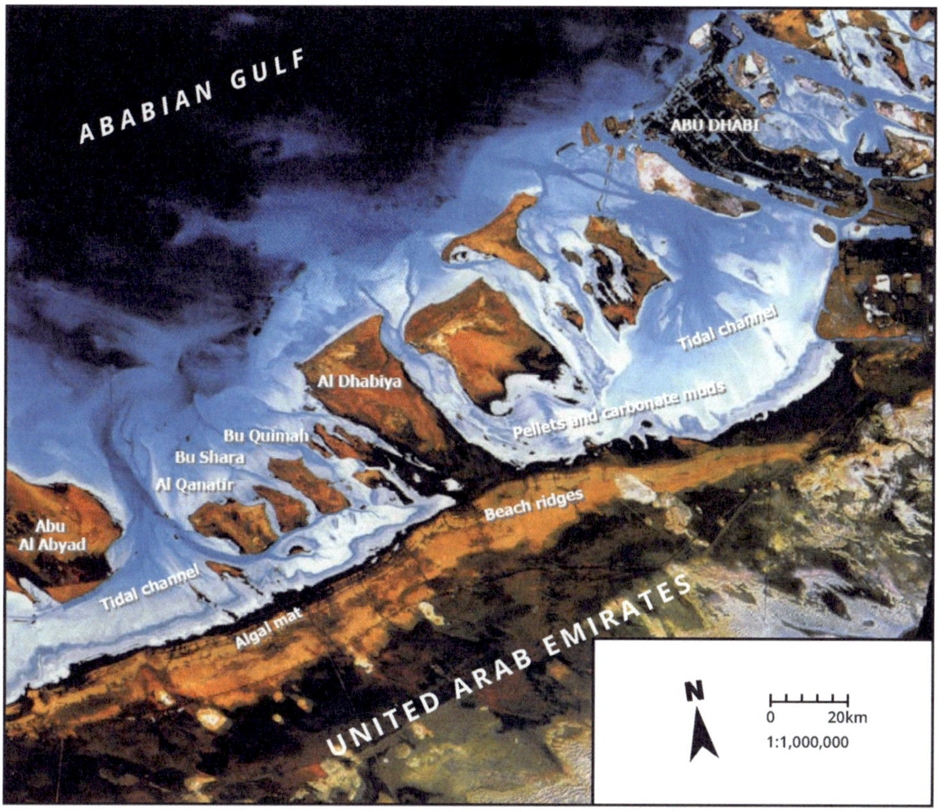

Figure 4.
Landsat image showing tidal channels, pellets, and carbonate mud between the islands. Beach ridges, algal mat, and coastal sabkha toward the coast [11].

These variations in the surface morphology are due to topographic or tidal range levels [12] and have been used by Butler et al. [7] to differentiate the upper salt flat (upper supratidal), the mid-salt flat (mid-supratidal), the gypsum mush (lower supratidal), and the algal mat (upper intertidal) facies, respectively. This sequence is representative also of the Al Qanatir field area (**Figure 2**), although notable areal variations occur due to disruption of the sabkha slope by tidal channel depressions and by elevated hardground tepee structures. The cyanobacterial material is preserved as peats beneath the supratidal sabkha, and they have a high potential for being preserved in the geologic record [4].

4.1.1 The upper salt flat facies

The upper salt flat facies are defined in this field area primarily by a thick upper sequence of anhydrite beneath a polygonal cracked surface crust of halite (**Figure 5A**). The halite polygons commonly overly, in descending order, a sequence of storm wash over carbonates, anhydrite, minor gypsum, algal peat, and medium-to-very coarse carbonate sand (**Figure 5B** and **C**), although all but the carbonate sand facies may be absent. Large (up to 6 cm) gypsum crystals are common within or immediately below the lowest algal peat layer, often preserving original algal laminations, and the surficial halite is thin less than 3 mm.

4.1.2 The mid-salt flat facies

The mid-salt flat is covered with a convoluted crust of halite (**Figure 5D**), which overlies a thin gypsum-replacive anhydrite that may be interlayered with storm wash over carbonate sediment, a thicker layer of gypsum, a well-developed algal peat, and again the thick sequence of carbonate sand (**Figure 5E**). The algal peat commonly displays desiccation polygons on the bedding plane, and hardgrounds, cemented by carbonates or gypsum, are common in the carbonate sand.

4.1.3 The gypsum mush facies

Gypsum actively precipitates at a depth of a centimeter or less beneath the surface sediment in the lower supratidal zone and beneath crenulated mats of algae in the upper intertidal zone. The sediment is moist, preventing formation of anhydrite except where dewatering of the sediment occurs, especially where footsteps and vehicle tracks have compressed the sediment. The surface carbonate washed over sediment may contain halite, but it does not form a crust at the surface (**Figure 5E**). The crystal in the gypsum is variable in size, from a few millimeters to a centimeter in diameter [13]. The crystals overlie an algal peat, commonly forming a transition zone of organic rich gypsum layers. Below the peat are the ubiquitous hardgrounds and unconsolidated carbonate sands.

4.1.4 The algal facies

The algal mats (or more appropriately, microbial mats, [14]) of Abu Dhabi have been described by and subdivided into groups based upon gross morphology of the mats [15]. The morphological variations represent differences in microbial species that form the mats, or modifications due to environmental conditions, primarily a result of the relative amounts of air exposure; thus, mat types can be directly correlated to the topographic level of the flat (**Figure 5F** and **G**).

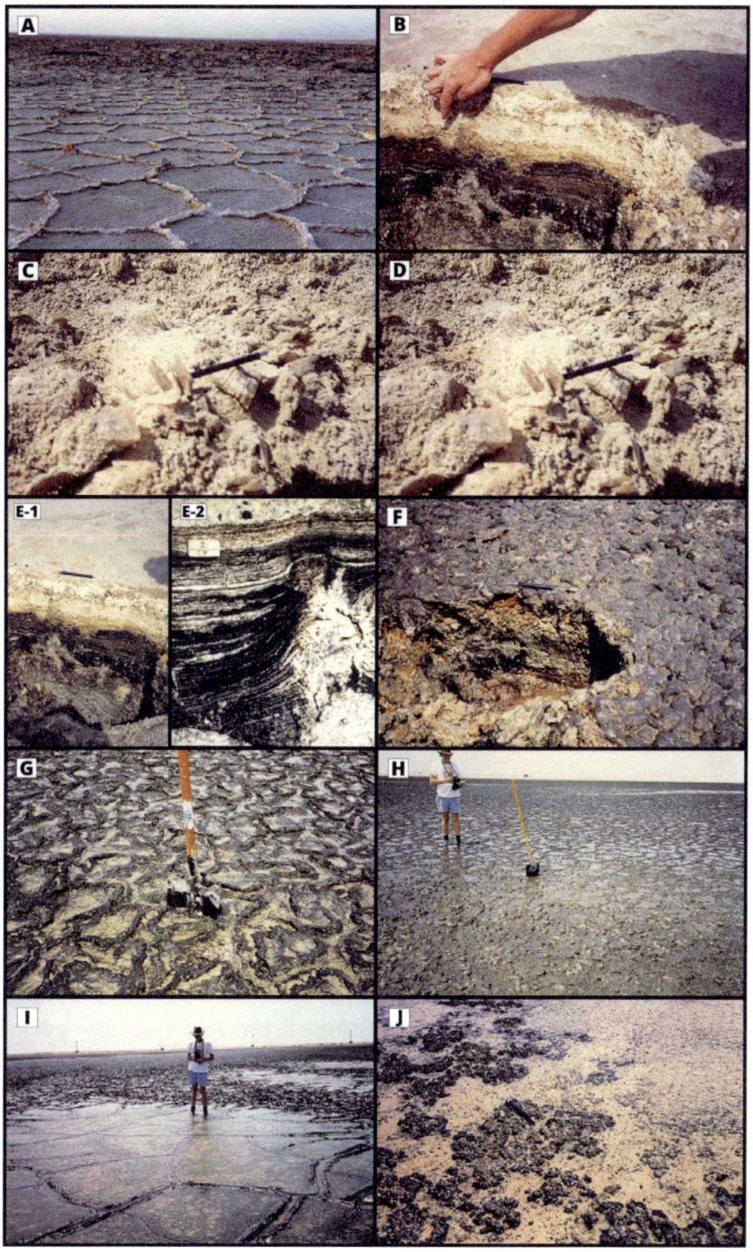

Figure 5.
Field photos of Holocene carbonates and evaporites of western Jazirat Al Qanatir: A. Polygonal-cracked surface crust of halite. B. Anhydrite nodules, diapiric anhydrite above gypsum mush and cyanobacterial mat. C. Gypsum rosettes. D. Sand-rich ridges of halite with a tepee surface. E. Gypsum much (with organic-rich gypsum crystals) above cyanobacterial peat and carbonate sand (E1). Detailed structure of algal mat showing deformation structures due to gas escape or burrowing, with banded anhydrite: some of them are enterolithic with complex ptygmatic folding and gypsum mush that occurs underlies the anhydrite sequence (E2). F. Crenulated or crinkle mat of the upper intertidal zone with gypsum mush and cyanobacterial peat below. G. Polygonal mat showing the upturned edges and discoloration toward the center of the mat. Quite often this is a pinkish color. H. Polygonal mat with the characteristic upturned edges in mid to upper intertidal zone. These pinnacle mats are abundant in areas with low tide exposure and occur as ripples caused by waves near shore. I. Large cyanobacterial mat polygons, which reach a few meters in diameter. J. The cinder-like cyanobacterial mat marks the seaward margin of the algal flats.

Several types and modifications are evident in the Al Qanatir samples. The mammillate [14] or cinder mat [15] inhabits the lowermost intertidal zone and is the first mat type in the vertical and lateral sequence of mat types in the mat facies. In the mid-intertidal zone, a smooth, continuous, and generally polygonally desiccated, flat mat occupies several sublevels and displays several morphologic modifications [13]. This mat is layered and has a beige to pink surface color, although variations occur that are dark greenish black. Within tidal creeks, the flat mat accumulates into thick sequences of peat and desiccates at the surface into large polygons. The edges of the polygons upturn and on a small scale provide a niche for pinnacle mat–type algae (**Figure 5H**). The pinnacles of the pinnacle mat are abundant in areas with low tide exposure and occur with several environmental modifications near the shore along transect. The upper intertidal zone, which is flooded during middle-to-high tide, is dominantly covered with the crenulated or crinkle zone mat ([15]; convoluted mat [14], under which the gypsum facies begin to develop (**Figure 5F**). The crenulated mat is a leathery, wrinkled mat, black on air-exposed upper surfaces of the folds, and commonly retains the pinkish-beige color on lower less exposed surfaces [13]. The fold trap air and gases are expelled when walked upon. A common morphological variation in the texture of these mats includes the formation of "tufts" on the upper more aerated surface of the crenulated mat, usually in the mid-intertidal zone, along transitions of the pinnacle or polygonal mats with the crenulated mat. In the uppermost intertidal zone, the crenulated mat completely desiccates and shrivels to a dried crust during low tide (**Figure 5I**). There is no significant accumulation of microbial material in this zone [11, 13]. The mammillated or cinder mat inhabits the lower most intertidal zone and is the first mat type in the vertical and lateral sequence in the mat facies (**Figure 5J**).

4.2 Preservation potential of cyanobacterial mats

Abu Dhabi algal mats have assessed the relative preservation potential of mat types based on the preservation of gross structures [16]. The preservation of the mat is a function of the subsurface extent (thickness and distribution) of the mat. The buried mat should look like its living counterpart on the surface. Geometric associations of the mat with inorganic sediment, such as well-formed laminae and desiccation polygons of the flat mat, are easily recognized. The flat mat can be recognized in the subsurface. A study of microstructures can delineate subtle diagenetic changes in the mat as well as the degree of preservation of the cyanobacterial constituents of the mat [4]. The amorphous organic material (sheath material and probably cell fluids) may degrade in a manner that destroys depositional microstructures.

Coccoid species (or at least the cell walls of these organisms) are relatively well preserved, regardless of the mat type in which they are found. Literature citing fossil examples of cyanobacterial material emphasize this point, since many of the examples are coccoid species similar (or perhaps identical) to those collected from coastal areas of Abu Dhabi (at Jazirat Al Qanatir) [17–19]. The dissolution of inorganic components (e.g., pods of micritic carbonates) and subsequent compaction of the mat may alter the depositional structure of the mat [4].

5. Laboratory results

Petrographic analyses from oriented microtome sections reveal the highly microlaminated structure of the algal mat sediments. These laminated sediments consist of alternating algal/bacterial layers and carbonate/evaporite layers (**Figure 4**). Filamentous blue-green algae are readily apparent. Almost all organic material was found to be of algal or bacterial origin and exhibited strong fluorescence (in the yellow green to light yellow range, typical of immature sediments). Semi-quantitative mineral analyses from x-ray diffraction are representative of evaporite sediments (**Table 1**). All samples contain significant amounts of halite. Most of the samples are dominated by carbonate minerals (aragonite, calcite, dolomite, and siderite). A few are dominated by sulfate minerals (gypsum and anhydrite). Small amounts of silicates (quartz and clay) are probably from onshore windblown particles carried to the sabkha.

The evolution of the interstitial waters within the sediments of the coastal sabkha has increased in salinities from the lagoonal edge to the inner parts of the sabkha and resulted in a series of mineralogical changes in which the associated organic material was involved and affected. The formation of gypsum and anhydrite both within the sediments and in the algal mats due to increase in salinity from lagoonal edge to thinner parts of the sabkha. The development of dolomite (secondary origin) since the initial sediments is composed of aragonite (ranges from less than 5% to more than 45%). The percentage of dolomite increases from the lagoonal margin of the sabkha toward its interior, as a function of increasing Mg/Ca ratio due to the precipitation

Sample ID	QZ	CA	DO	PY	SD	ANH	GYP	HA	Arag
1	0	0	0	0	5	0	33	36	26
2	0	0	0	5	0	33	40	22	0
3	6	0	0	0	0	0	31	42	21
4	2	4	0	5	0	0	1	47	40
5	4	6	16	0	0	0	1	46	25
6	4	5	3	4	0	0	0	40	42
7	2	0	0	0	5	0	41	27	26
8	1	6	0	6	0	0	8	29	50
9	0	0	0	0	8	0	62	25	4
10	3	9	5	0	0	0	0	50	34
11	3	15	0	4	0	0	12	58	7
12	3	0	0	4	2	0	12	47	33
13	9	3	0	0	0	0	0	44	35
14	3	9	2	5	0	0	0	29	47
15	8	11	2	5	0	0	0	28	45
16	4	10	2	7	0	0	0	28	45
17	3	7	9	3	0	0	0	46	28
18	1	7	0	0	3	0	25	44	16
19	1	5	1	1	2	0	18	40	14

QZ = Quartz; PY = Pyrite; GYP = Gypsum; CA = Calcite; SD = Siderite; HA = Halite; DO = Dolomite; ANH = Anhydrite; Arag = Aragonite.

Table 1.
Whole rock semi-quantitative x-ray diffraction mineralogic analysis (sabkha near surface samples from south Al Qanatir, western Abu Dhabi, UAE).

of calcium sulfate minerals. The transformation of aragonite to calcite, probably due to the progressive loss of magnesium from the interstitial waters, or to the effects of meteoric waters as mentioned by many authors [8, 20].

Microscopic evidence shows that the solid organic matter is mostly highly fluorescent algal material, which is relatively free of microbial maceration. This suggests that the organic matter is well enough preserved to become a source rock. Interpretation of the pyrolysis data indicates that little similarity exists between these algal samples and terrestrial peat samples. The most noticeable difference is the lack of lignin pyrolysis

Sample ID	OC (%)	S1	S2	S1 + S2	S3	HI (S2/OC)	OI (S3/OC)	KTR (S1/S1 + S2)	TMAX (OC)
1	0.58	—	—	—	—	—	—	—	—
2	0.91	—	—	—	—	—	—	—	—
3	1.24	0.98	4.89	5.87	3.76	395	303	0.17	433
4	2.84	2.53	10.89	13.42	7.24	383	254	0.19	420
5	2.14	1.54	10.21	11.75	6.26	477	292	0.13	436
6	1.18	0.69	4.67	5.36	5.05	394	426	0.13	435
7	2.74	4.59	13.94	18.53	8.24	508	300	0.25	422
8	2.33	3.46	11.9	15.36	6.21	510	266	0.23	418
9	1.98	2.35	10.7	13.05	6.59	541	333	0.18	426
10	2.19	1.7	1064	12.34	7.04	484	320	0.14	439
11	1.34	1.17	7.22	8.39	6.76	537	503	0.14	442
12	1.28	1.01	5.09	6.1	3.94	396	303	0.17	431
13	1.62	3.25	6.98	10.23	5.01	430	308	0.32	426
14	6.88	20.85	26.78	47.63	11.24	389	163	0.44	U
15	5.01	7.57	23.05	30.62	9.96	459	198	0.25	428
16	2.06	1.96	10.53	12.49	5.46	510	264	0.16	434
17	8.4	22.18	33.04	55.22	13.82	393	164	0.4	U
18	1.84	2.5	10.38	12.86	6.76	564	367	0.2	430
19	2.01	2.71	11.99	14.7	6.26	567	311	0.18	437
20	0.46	—	—	—	—	—	—	—	—

OC = Organic carbon (wt. %).
S1 = generated hydrocarbons from pyrolysis of (mg/g of rock) kerogen hydrocarbon generation potential of the rock.
S1 + S2 = total hydrocarbon potential of the rock (mg/g of rock).
S3 = CO^2 generated from kerogen pyrolysis (mg/g of rock).
HI = hydrogen index (mg generated hydrocarbon/g of organic carbon).
OI = oxygen index (mg CO^2/g of organic carbon).
KTR = kerogen transformation ratio proportion of distillable hydrocarbons to total hydrocarbon potential as a maturation index when samples are contamination free.
TMAX = pyrolysis temperature at which a maximum yield of generated hydrocarbons occurs. TMAX increases with increasing maturation.
I = hydrocarbon yield (S2 and/or S2) is too low for reliable interpretations.
U = TMAX is less than 400 °C and is not representative of kerogen transformation.

Table 2.
Organic carbon and Rock-Eval pyrolysis data (near surface cores from sabkha south Al Qanatir), western Abu Dhabi, UAE.

products in the algal samples, which is indicative of the lack of higher cellular structure in the algae. Another significant difference is an abundance of nitrogen-containing compounds in the algae.

Contrary to terrestrial peats, these sediments are rich in indoles and substituted indoles. The origin of these indole pyrolysis products is not known, but they may be derived from the growth regulator indoleacetic acid. Phenolic compounds are also abundant in the algae. Phenol, methyl-, dimethyl-, methylethyl-, and ethyl-phenols, along with dihydroxy phenols have all been identified as pyrolysis compounds. Large quantities of phenolic materials can be released by certain species of algae. The polysaccharide pyrolysis products from the algal samples are similar to those from terrestrial peats. Furans, furanones, and furaldehyde, among other products, are present in both sets of samples and most likely represent a very basic form of cellular morphology common to most forms of living organisms (as described in [21, 22]). The remainder of the pyrolysis products represent unclassified lipid fractions, with then-alkanes (C^{10}-C^{20}) being observed along with other unidentified lipid fragments. Results from geochemical analysis are shown in **Table 2**. Interpretation of these chemical analysis included whole rock pyrolysis, standard organic carbon analysis (TOC), and elemental analysis (ash-free), which are presented in **Figures 6–10**.

TOC and Rock-Eval pyrolysis show a wide range of organic enrichment (0.46 to 8.40% TOC) with moderately high hydrogen indices (389–597), typical of marine Type II kerogens (**Figure 6**). However, oxygen indices (OI) are also very high (163–503). Other analyses of mats collected from western Abu Dhabi by Kenig et al. [5] show various values (0.5–2.7% TOC and HI from 510 to 675). In ancient sediments, high oxygen indices values usually indicate an oxidized kerogen, due to weathering, an oxidative depositional environment, or benthic biodegradation. For recent evaporite sediments containing high proportions of aragonite, other contributing

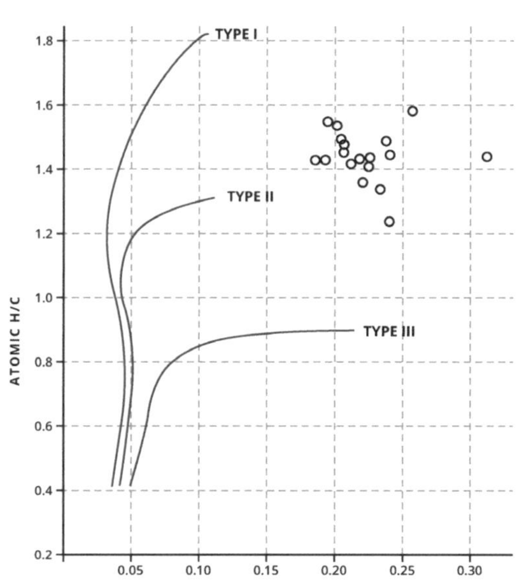

Figure 6.
Modified van Krevelen diagram showing H/C to O/C relationship from elemental analysis of algal mat samples from western Abu Dhabi [4].

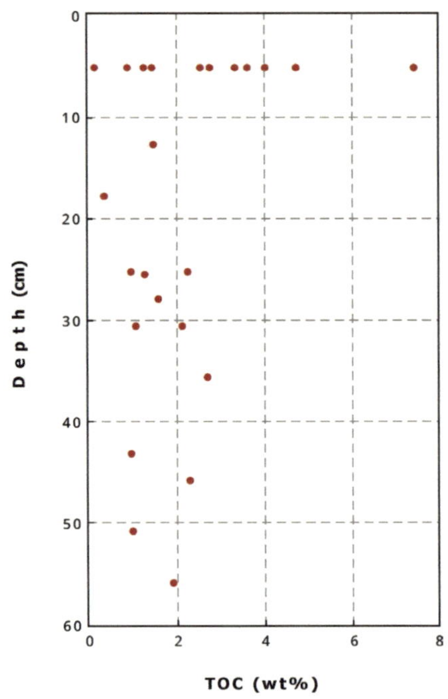

Figure 7.
Relationship between depth (cm) and TOC (wt.%) from algal mat samples from Abu Dhabi.

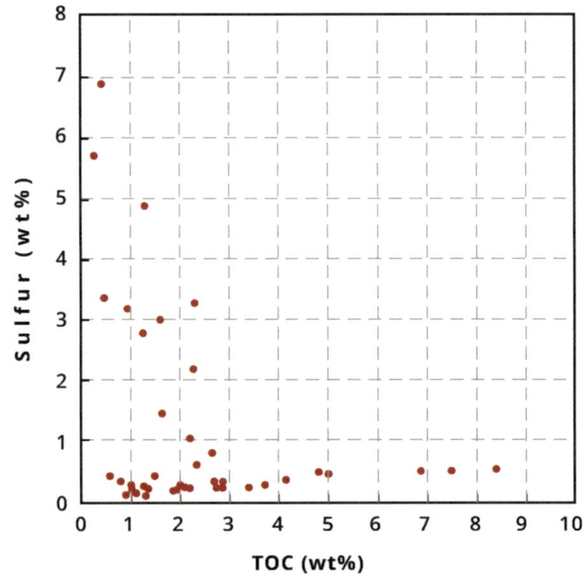

Figure 8.
Relationship between sulfur (wt.%) and TOC (wt.%) from algal mat samples from Abu Dhabi.

factors must be considered. Since aragonite is the less stable precursor to calcite, it is possible that a proportion of aragonite is breaking down under pyrolytic conditions and producing enough carbon dioxide to interfere with the carbon dioxide produced from organic thermal degradation. Recently deposited organic matter tends to have a

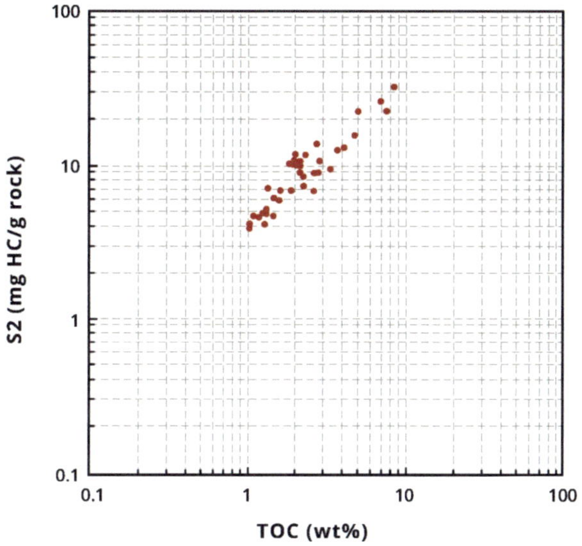

Figure 9.
Relationship between TOC (wt.%) and S2 pyrolysis kerogen from algal mat samples from Abu Dhabi.

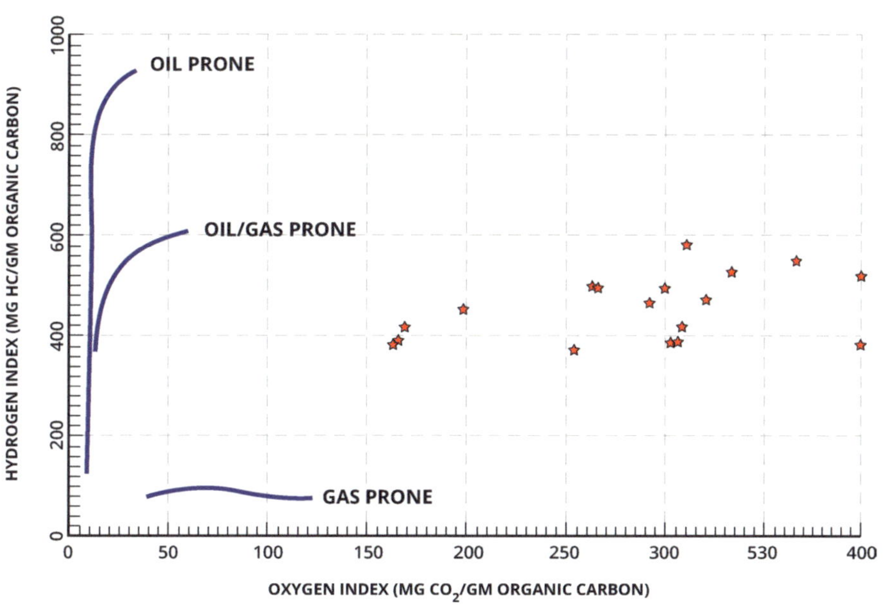

Figure 10.
Relationship between hydrogen index (HI) and oxygen index (OI) from algal mat samples from Abu Dhabi.

high degree of functionality (carboxyl, carbonyl and hydroxyl groups, as well as ether linkages) that contributes to the oxygen index.

Elemental analysis (C, H, N, O, S) of isolated solid organic material showed atomic hydrogen to carbon (H/C) ratios (1.20–1.54) that fall directly on the Type II evolutionary pathway as shown in a modified van Krevelen diagram [23]. Atomic oxygen to carbon ratios (O/C) (0.19–0.32) indicate that the sedimentary organic matter is not as oxygen rich as many of the oxygen index pyrolysis results initially suggested.

This further supports the idea that aragonite does interfere with the pyrolysis oxygen index results. Carbon isotope measurements of isolated solid organic material showed $^{13}C/^{12}C$ ratios (−8.41 to −10.78%°) that are consistent with modern aquatic plants (−6 to −19%° relative to PDB). Also, the narrow range of carbon isotope ratios indicates that the sedimentary organic matter is similar over the sampled area.

Microscopic evidence shows that the solid organic matter is mostly highly fluorescent algal material that is relatively free of microbial maceration. Total organic carbon (TOC) values and Rock-Eval pyrolysis yields are high in some samples (in this study and in [3, 5]); this suggests that the organic matter is well enough preserved to become a source rock. Although TOC and Rock-Eval techniques were developed to assess petroleum source rock potential for Jurassic Diyab Formation in UAE that have at least undergone organic diagenesis [24], analyses of these recent cyanobacterial sediments indicate significant organic enrichment, with potential to generate hydrocarbons upon thermal degradation. These results represent significant evidence that rocks of this type can be potential petroleum sources.

6. Significance to petroleum exploration

Recent algal mats or cyanobacterial mats occur in surface and subsurface accumulations of sabkha sediment and form bands along the coastline of Abu Dhabi. This association of concentrations of organic matter interbedded with carbonate-evaporite is uncommon in the Holocene but is believed to have been more common in the geologic past and may represent potential source rocks. Now, the petroleum potential of organic material in this type of environment is of interest since many of the world's largest oil fields occur in ancient carbonate-evaporite sequences similar to that of Abu Dhabi coastal areas. In assessing the petroleum potential of these cyanobacterial

Field name	Formation	Age	Location
Puckett	Ellenberger	Ordovician	Texas, USA
Cabin Creek & Pennel	Red River	Silurian	Montana, USA
Rainbow	Keg River	Middle Devonian	Alberta, Canada
Northwest Lisbon	Leadville	Mississippian-Pennsylvanian	Paradox Basin, USA
Little Knife	Mission Canyon	Mississippian	North Dakota, USA
Tarchaly, Rybaki & Sulecin	Zechstein	Early Permian	Poland
Qatif	Arab	Late Jurassic	Saudi Arabia
Ghawar	Arab	Late Jurassic	Saudi Arabia
Umm Shaif	Arab	Late Jurassic	UAE
Dukhan	Arab	Late Jurassic	Qatar
Chatom	Smackover	Late Jurassic	Arkansas, USA
Mt. Vernon	Smackover	Late Jurassic	Arkansas, USA
Sunniland	Sunniland	Early Cretaceous	Florida, USA
Gachsaran & Bibi Hakimeh	Asmari	Oligocene-Early Miocene	Iran

Table 3.
Some carbonate hydrocarbon reserves associated with tidal flat evaporates and algal flats [10, 25].

accumulations, the preservation potential as well as chemistry of the organics must be considered.

Shoaling-upward sequences of carbonates and evaporites are common hydrocarbon reservoirs. Examples include grain carbonates of the Upper Jurassic Arab Formation and its associated Hith Anhydrite cap rock of eastern Arabian Basin oil and gas fields [6]. Together, these form the world's largest oil fields (examples cited in **Table 3**). In the United States, similar reservoirs are associated with major oil fields in the Permian carbonates of the Central Basin platform and on the Northwest shelf of the Permian Basin of Texas and New Mexico (where the upper shelf carbonates interfinger with up dip evaporites and clastics). This association also occurs within fields of the Williston Basin associated with the Mississippian Madison Limestone/Charles evaporite couplets, and with the Ordovician Red River Formation [25]. Similar rocks can also be found in the western Canadian Basin, where the Devonian Winnipegosis shoaling-upward carbonates and evaporites are associated with sequences that are similar to those of UAE [26]. Since hydrocarbon reservoirs related to evaporite/carbonate cycles are common in the geological record, the hypothesis presented here is that the algal sediments that occur in modern settings may also have been present in similar ancient sequences and consequently could account for the occurrence of petroleum in these sequences [2, 9].

7. Conclusions

Laboratory studies indicate that concentrations of organics in these algal deposits based on TOC analyses show some similarities in ancient source rocks in some fields in Middle East and North America. Hydrocarbon generation potential based on Rock-Eval pyrolysis (HI values) suggest that the organic matter can generate oil and gas upon thermal maturation. Both Rock-Eval pyrolysis and elemental analysis show that organic matter follows the Type II kerogen evolution pathway. Microscopic evidence shows that the solid organic matter is mostly highly fluorescent algal material that is relatively free of microbial maceration. This suggests that the organic matter is well enough preserved to become a source rock.

Acknowledgements

I gratefully acknowledge my colleague, the late Chris Kendall, whose insights and dedication were invaluable to this work. Part of this work can be seen at SEPM website http://sepmstrata.org/page.aspx?3=&pageid=189

Author details

Abdulrahman Alsharhan[1*] and Christopher Kendall[2†]

1 Middle East Geological and Environmental Establishment, Dubai, UAE

2 Department of Geology, University of South Carolina, Columbia, SC, United States America

*Address all correspondence to: sharhana@emirates.net.ae

† Christopher Kendall as deceased. on 15-10-2024.

References

[1] Kendall CGSC, Alsharhan AS. Holocene geomorphology and recent carbonate-evaporite sedimentation of the coastal region of Abu Dhabi, United Arab Emirates. In: Kendall CGSC, Alsharhan AS, editors. International Association of Sedimentology. Special Publication-43. UK: Wiley-Blackwell; 2011. pp. 45-88

[2] Alsharhan AS, Kendall CGSC. Holocene carbonate-evaporites of Abu Dhabi, and their Jurassic ancient analogs. In: Barth HJ, Boer B, editors. Sabkha Ecosystems. London: Kluwer Academic Publisher; 2002. pp. 187-202

[3] Cardoso JN, Watts CD, Maxwell JR, Goodfellow R, Eglinton G, Golubic S. A biogeochemical study of the Abu Dhabi algal mats: A simplified ecosystem. Chemical Geology. 1978;**23**:273-291

[4] Kendall CGSC, Alsharhan AS, Cohen A. The Holocene tidal flat complex of the Arabian Gulf coast of Abu Dhabi. In: Barth H-J, Boer B, editors. Sabkha Ecosystems - The Arabian Peninsula and Adjacent Countries. The Netherlands: Kluwer Academic Publishers; 2002. pp. 21-35

[5] Kenig F, Huc AY, Purser BH, Oudin JL. Sedimentation, distribution and diagenesis of organic matter in a recent carbonate environment Abu Dhabi, UAE. Organic Geochemistry. 1990;**16**:735-745

[6] Alsharhan AS, Kendall CGSC. Depositional setting of the upper Jurassic Hith anhydrite of the Arabian Gulf: An analog to holocene evaporites of the United Arab Emirates and Lake MacLeod of Western Australia. AAPG Bulletin. 1994;**78**:1075-1096

[7] Butler GP, Harris PM, Kendall CGSC. Recent evaporites from the Abu Dhabi coastal flats. In: Depositional and Diagenetic Spectra of Evaporites. Calgary: SEPM Core Workshop-3; 1982. pp. 33-64

[8] Bush P. Some aspects of the diagenetic history of the sabkha in Abu Dhabi, Persian Gulf. In: Purser BH, editor. The Persian Gulf: Holocene Carbonate Sedimentation and Diagenesis in a Shallow Epicontinental Sea. Berlin: Springer-Verlag; 1973. pp. 395-407

[9] Alsharhan AS, Kendall CGSC. Holocene coastal carbonates and evaporites of the southern Arabian Gulf and their ancient analogues. Earth Science Review. 2003;**61**:191-243

[10] Alsharhan AS, Kendall CGSC. Introduction to quaternary carbonate and evaporite sedimentary facies and their ancient analogues. In: Kendall CGSC, Alsharhan AS, editors. Quaternary Carbonate and Evaporite Sedimentary Facies and their Ancient Analogues - A Tribute to Douglas J. Shearman. International Association of Sedimentology (IAS), Special Publication-43. UK: Wiley-Blackwell; 2011. pp. 1-10

[11] Alsharhan AS, Kendall CGSC. Interpretations of Holocene carbonate-evaporites of coastal and inland sabkhas of Abu Dhabi (United Arab Emirates) from Landsat satellite images and field survey. In: Gul B et al., editors. Sabkha Ecosystems, Tasks for Vegetation Science VI. Switzerland: Springer Nature; 2018. pp. 151-187

[12] Butler GP. Modern evaporative deposition and geochemistry of coexisting brines, the sabkha: Trucial coast. Arabian Gulf: Journal of Sedimentary Petrology. 1969;**39**(1):70-89

[13] Kendall CGSC, Alsharhan AS, Whittle GL. The flood re-charge sabkha model supported by recent inversions of anhydrite to gypsum in the UAE sabkhas. In: Alsharhan AS, Glennie KW, Whittle GL, Kendall CGSC, editors. Quaternary Deserts and Climatic Change. Rotterdam: A.A. Balkema; 1998. pp. 29-42

[14] Golubic S. Microbial mats of Abu Dhabi. In: Margulis L, Olendzenski L, editors. Environmental Evolution; Effects of the Origin and Evolution of Life on Planet Earth. Cambridge, USA: MIT Press; 1992. pp. 103-130

[15] Kendall CGSC, d'ESkipwith PA. Recent algal mats of a Persian Gulf lagoon. Journal of Sedimentary Petroleum. 1968;**38**:1040-1058

[16] Park RK. The preservation potential of some recent stromatolites. Sedimentology. 1977;**24**:485-506

[17] Awramik SM. Ancient stromatolites and microbial mats. In: Microbial Mats Stromatolites. New York: Alan R. Liss, Inc. Press; 1984. pp. 1-22

[18] Golubic S, Barghoom ES. Interpretation of microbial fossils with special reference to the Precambrian. In: Flugel E, editor. Fossil Algae. Berlin: Springer-Verlag; 1977. pp. 1-14

[19] Golubic S, Yan Z. Phycological expertise in geological applications. Hydrobiologia. 1985;**123**:193-198

[20] Butler GP. Recent gypsum and anhydrite of the Abu Dhabi sabkha, Trucial coast: An alternative explanation of origin. In: Third Symposium on Salt. Vol. 1. Ohio, USA: Northern Ohio Geological Society; 1970. pp. 120-152

[21] Kenig F, Damste JSS, Dalen ACK, Rijpstra WIC, Huc AY, de Leeuw JW. Occurrence and origin of mono-, di-, and trimethylalkanes in cyanobacterial mats from Abu Dhabi, United Arab Emirates. Geochimeca et Cosmochimica Acta. 1995;**59**(14):2999-3015

[22] Scherf AK, Rullkotter J. Biogeochemistry of high salinity microbial mats- part 1: Lipid composition of microbial mats across intertidal flats of Abu Dhabi, United Arab Emirates. Organic Geochemistry. 2009;**40**(9):1018-1028

[23] Tissot BP, Welte DH. Petroleum Formation and Occurrence. 2nd ed. Berlin: Springer-Verlag; 1984. 520 p

[24] Al-Suwaidi AS, Taher AK, Alsharhan AS, Salah MG. Stratigraphy and geochemistry of upper Jurassic Diyab formation, Abu Dhabi, UAE. In: Alsharhan AS, Scott R, editors. Middle East Models of Jurassic/Cretaceous Carbonate Systems. SEPM Special Publication-69. Oklahoma, USA: Society of Sedimentary Geology, Tulsa; 2000. pp. 249-272

[25] Roehl PO, Choquette PW, editors. Carbonate Petroleum Reservoirs. New York: Springer Verlag; 1985. 650 p

[26] Shearman DJ, Fuller JG. Anhydrite diagenesis, calcitization and organic laminites, Winnipegosis formation, middle Devonian, Saskatchewan, Canada. Canadian Petroleum Geology Bulletin. 1969;**17**:496-525